CREATED FROM ANIMALS

CREATED
FROM
ANIMALS

THE MORAL IMPLICATIONS
OF DARWINISM

JAMES RACHELS

OXFORD
UNIVERSITY PRESS

OXFORD
UNIVERSITY PRESS

Oxford   New York
Athens   Auckland   Bangkok   Bogotá   Buenos Aires   Calcutta
Cape Town   Chennai   Dar es Salaam   Delhi   Florence   Hong Kong   Istanbul
Karachi   Kuala Lumpur   Madrid   Melbourne   Mexico City   Mumbai
Nairobi   Paris   São Paulo   Singapore   Taipei   Tokyo   Toronto   Warsaw

and associated companies in
Berlin   Ibadan

Copyright © 1990 by James Rachels

First published by Oxford University Press, Inc., 1990

First issued as an Oxford University Press paperback, 1999
198 Madison Avenue, New York, New York 10016

Oxford is a registered trademark of Oxford University Press

Library of Congress Cataloging-in-Publication Data
Rachels, James, 1941-
Created from animals : the moral implications of Darwinism / James Rachels.
p.   cm.
Includes bibliographical references.
ISBN 0-19-217775-3 (cloth)
ISBN 0-19-286129-8 (pbk.)
1. Evolution--Moral and ethical aspects. I. Title
B818.R32 1990   171'.7--dc20   89.38670

1 3 5 7 9 10 8 6 4 2

Printed in the United States of America

CONTENTS

INTRODUCTION

> Man in his arrogance thinks himself a great work worthy
> the interposition of a deity. More humble and I think
> truer to consider him created from animals.

DARWIN wrote these words in 1838, twenty-one years before he was to publish *The Origin of Species*. He would go on to support this idea with overwhelming evidence, and in doing so he would bring about a profound change in our conception of ourselves. After Darwin, we can no longer think of ourselves as occupying a special place in creation—instead, we must realize that we are products of the same evolutionary forces, working blindly and without purpose, that shaped the rest of the animal kingdom. And this, it is commonly said, has deep philosophical significance.

The religious implications of Darwinism are often discussed. From the outset, churchmen have worried that evolution is incompatible with religion. Whether their concern is justified is still debated, and I will have a good bit to say about this. But Darwinism also poses a problem for traditional morality. Traditional morality, no less than traditional religion, assumes that man is 'a great work'. It grants to humans a moral status superior to that of any other creatures on earth. It regards human life, and only human life, as sacred, and it takes the love of mankind as its first and noblest virtue. What becomes of all this, if man is but a modified ape?

Curiously, philosophers have shown little interest in such questions. The proverbial 'man in the street' might believe that there are big philosophical lessons to be learned from Darwin—or big threats posed by Darwin—but by and large academic thinkers have not agreed. In the decades immediately following the publication of Darwin's theory, some philosophers did have a lot to say about it. Then it was fashionable to think that Darwinism had deep implications for everything. But this interest quickly waned. If we examine the most influential works of philosophy written in the twentieth century, we find few references to Darwin. His theory is discussed, of course, in works devoted narrowly to the philosophy of science. But in philosophical works of more general interest, and particularly in books about ethics, it is largely ignored. When the subject is broached, it is usually to explain that Darwinism does not have some implication it is popularly thought to have. The philosophers seem to agree with

Wittgenstein's assessment: 'The Darwinian theory', said Wittgenstein, 'has no more to do with philosophy than any other hypothesis of natural science.'

Why have philosophers, with a few exceptions, been so indifferent to Darwin? Partly it may be a reaction to the absurdity of claims that were once made. When he first read the *Origin*, Karl Marx declared that 'Darwin's book is very important and serves me as a basis in natural selection for the class struggle in history.' Later socialists made similar judgements, claiming to find in Darwin the 'scientific basis' of their political views. Meanwhile, capitalists were also claiming him: in the late nineteenth century the idea of 'the survival of the fittest' was invoked again and again to justify competitive economic systems. In 1900 the American industrialist Andrew Carnegie wrote that we must 'accept and welcome . . . great inequality; the concentration of business, industrial and commercial, in the hands of a few; and the law of competition between these, as being not only beneficial, but essential to the future progress of the race'. Why? Because capitalism alone 'ensures the survival of the fittest'. To make things even worse, Heinrich Himmler would later claim that Darwinism supported purging Europe of the 'unfit' Jews. Exasperation with such nonsense might very well provoke a reaction like Wittgenstein's. In the face of all this, it is tempting simply to throw up one's hands and say: Darwin's theory is about biology, *not* politics or economics or ethics or religion or anything else.

Educated people might resist the idea that Darwinism has moral implications for still another reason. Many people today think that Darwinism is *contrary* to true morality, and they reject it for that reason. Most of the current resistance to Darwinism seems to be at least partially motivated by this thought. In the United States, there are those who would like to ban the teaching of evolution in public schools; to stir up public feeling, they point to its supposedly obnoxious implications for religion and morality. The argument is depressingly familiar. The idea that Darwinism undermines traditional values has now been used so often as a reason for objecting to the theory that scientifically minded people might naturally think it is nothing but an ignorant notion, to be rejected out of hand.

The leading defenders of evolution take just this position—they insist that their theory can pose no threat to morality or religion because their theory *has no* implications for morality or religion. Stephen Jay Gould, one of the foremost contemporary defenders of Darwinism, and certainly our most effective writer on the subject, responds to the right-wing challenge by deploring 'the silly dichotomy of science versus religion', and by assuring his readers that 'While I'm not a conventional believer, I don't consider myself irreligious.' He goes on to urge that there is no conflict between Darwinism and old-fashioned values, or, for that matter, any kind of values at all:

What challenge can the facts of nature pose to our own decisions about the moral value of our lives? We are what we are, but we interpret the meaning of our heritage as we choose. Science can no more answer the questions of how we ought to live than religion can decree the age of the earth.

Thus, as the debate goes on, only two positions seem possible: the fundamentalist view that Darwinism undermines traditional values, and so must be rejected; and the evolutionist reply that Darwinism poses no threat to traditional values. When the lines are drawn this way, it is difficult to take seriously the possibility that Darwinism might have moral consequences—especially the notion that Darwinism undermines traditional morality—without seeming to side with evolution's enemies. The upshot is that, in learned circles, it is commonly taken to be a sign of enlightenment to believe that Darwinism has no implications for ethics. Lost in the fog is the possibility of a third alternative: that Darwinism is incompatible with traditional morality, and so provides reason for rejecting that morality and replacing it with something better.

But there is a deeper, more principled reason for scepticism about finding moral lessons in Darwinism, hinted at in Gould's argument. That is the old problem of the relation of fact and value, of 'is' and 'ought'. We cannot, as a general rule, validly derive conclusions about what *ought* to be the case from premisses about what *is* the case. Darwin's theory, if it is correct, concerns matters of fact. It tells us what is the case, with respect to the evolution of species. Therefore, strictly speaking, no conclusion follows from it regarding any matter of value. It does not follow, merely because we are kin to the apes, that we ought to think less of ourselves, that our lives are less important, or that human beings are 'merely' one kind of animal among others. Nor does it follow that the main tenets of religion are false. As has often been observed, natural selection could be the means by which God has chosen to make man. If so, man could still be regarded as the divinely blessed crown of creation.

Nevertheless, nagging thoughts remain. Can it really be true that Darwinism, which overturns all our former ideas about man and nature, has no unsettling consequences? Traditional morality is based, in part, on the idea that human life has a special value and worth. If we must give up our inflated conception of ourselves, and our picture of the world as made exclusively for our habitation, will we not have to give up, at the same time, those elements of our morality which depend on such

conceptions? The feeling that Darwin's discovery undermines traditional religion, as well as some parts of traditional morality, will not go
away, despite the nice logical points about what follows from what, and
despite the fact that one might not want to side with evolution's enemies.
I believe this feeling is justified. There is a connection between Darwin's
theory and these larger matters, although the connection is more complicated than simple logical entailment.

I shall argue that Darwin's theory does undermine traditional values.
In particular, it undermines the traditional idea that human life has a
special, unique worth. Thus, although I am a Darwinian, I will be
defending a thesis that Darwin's friends have usually resisted. But I do
not assume, as Darwin's enemies have assumed, that this implication of
Darwinism is morally pernicious. I believe it is a positive and useful
result that should be welcomed, not resisted. Abandoning the idea that
human life has special importance does not leave us morally adrift; it only
suggests the need for a different and better anchor.

Darwin said that *The Origin of Species* was 'one long argument'. At the
risk of seeming presumptuous, I would like to say the same thing about
the present book, that it also elaborates one long argument. The book
contains a good bit of intellectual history. This history is recounted,
partly to provide background, but also because I want to present my
philosophical argument in the context of the human events that made it
possible. Philosophical arguments are often presented ahistorically, as
abstract chains of reasoning whose logical validity is independent of
cultural context. There is nothing wrong with that way of writing;
indeed, it has been the norm among philosophers for most of this century. But in this book I have departed from this practice and have
included somewhat more historical material than is usual in a philosophy book that has an argument.

The argument may be summarized briefly:

1. Traditional morality depends on the idea that human beings are in
a special moral category: from a moral point of view, human life has a
special, unique value, while non-human life has relatively little value.
Thus the purpose of morality is conceived to be, primarily, the protection of human beings and their rights and interests. This is commonly referred to as the idea of human dignity. But this idea does not
exist in a logical vacuum. Traditionally it has been supported in two
ways: first, by the notion that man is made in the image of God, and
secondly, by the notion that man is a uniquely rational being.

2. Darwin's theory does not entail that the idea of human dignity is
false—to say that it does would violate the logical stricture against
deriving 'ought' from 'is'. Darwinism does, however, *undermine* the

traditional doctrine, in a sense that I will explain, by taking away its support. Darwinism undermines both the idea that man is made in the image of God and the idea that man is a uniquely rational being. Furthermore, if Darwinism is correct, it is unlikely that any other support for the idea of human dignity will be found. The idea of human dignity turns out, therefore, to be the moral effluvium of a discredited metaphysics.

3. To replace the doctrine of human dignity, I offer a different conception, moral individualism, which I argue is more in keeping with an evolutionary outlook. According to moral individualism, the bare fact that one is human entitles one to no special consideration. How an individual should be treated depends on his or her own particular characteristics, rather than on whether he or she is a member of some preferred group—even the 'group' of human beings. I offer various reasons for thinking this approach is morally sound, as well as reasons for thinking it is the natural view to take if one views the world from an evolutionary perspective.

4. Finally, abandoning the idea of human dignity, and adopting moral individualism in its place, has practical consequences. Human life will no longer be regarded with the kind of superstitious awe which it is accorded in traditional thought, and the lives of non-humans will no longer be a matter of indifference. This means that human life will, in a sense, be devalued, while the value granted to non-human life will be increased. A revised view of such matters as suicide and euthanasia, as well as a revised view of how we should treat animals, will result. I hope to show that reconstructing morality without the assumption of man's specialness leaves morality stronger and more rational. It leaves us with a better ethic concerning the treatment of both human and non-human animals.

There is one other theme I wish to pursue, about the scope of Darwin's work. As we shall see, Darwin himself had a good bit to say about morality and religion. But his remarks on these subjects are often ignored, or treated as only marginally interesting. The assumption seems to be that his views about morality and religion are independent of his strictly scientific project and have less value. Darwin himself, however, seems to have believed that all his thinking was of one piece. I shall argue that he was right: he may profitably be viewed as a systematic thinker whose views on all these subjects are closely related. Today almost everyone agrees that Darwin was a profound thinker. But I hope to show that he was a deeper thinker on a wider range of subjects than is commonly realized.

Darwin's Discovery

CHARLES DARWIN and Karl Marx, the two great revolutionaries of nineteenth-century thought, were almost exact contemporaries. Their deaths in 1882 and 1883 were separated by only a few months. For more than three decades, while writing their most important works, they lived sixteen miles apart, Marx in London and Darwin at his home in the Kentish village of Downe. But they never met, which is not surprising, considering how different were their lives and personalities. Marx spent much of his life in abject poverty, unable at times to feed his family. Because of his political activities, he was chased from country to country. Above all Marx was preoccupied with the great public events of his time: with industrialization, with political revolution, and with the vast changes he saw taking place in society. Darwin, by contrast, was oblivious to such matters. Born into a life of privilege, he would have a tranquil existence, rooted in one place, surrounded by servants and a loving family: the very picture of a nineteenth-century English gentleman. His revolution would be of a different sort.

DARWIN'S EARLY LIFE

When Darwin was born, on 12 February 1809, his family had high hopes. It was, after all, a noteworthy clan. Charles's father, Robert Waring Darwin II, a prosperous doctor in Shrewsbury, was himself the son of a distinguished father, the speculative evolutionist Erasmus Darwin. His mother, Susannah Wedgwood, who died when Charles was eight, was the daughter of Josiah Wedgwood, founder of the famous pottery. But Robert Darwin soon despaired that his son would ever amount to anything: 'You care for nothing but shooting, dogs, and rat-catching, and you will be a disgrace to yourself and all your family', the elder Darwin is said to have remarked. In this *Autobiography*, written privately for his family in 1876, Darwin himself remembered his youth as unpromising. His biographers have generally accepted this estimate.

Ronald W. Clark says flatly: 'Darwin had a youth unmarked by the slightest trace of genius.'

This judgement, however, seems altogether too harsh. Even as a boy, Darwin was unusual in his love of nature. Collecting and studying insects, especially beetles, was an all-absorbing passion. 'One day', he wrote in the *Autobiography*, 'on tearing off some old bark, I saw two rare beetles and seized one in each hand; then I saw a third and new kind, which I could not bear to lose, so that I popped the one which I held in my right hand into my mouth. Alas it ejected some intensely acrid fluid, which burnt my tongue so that I was forced to spit the beetle out, which was lost, as well as the third one.' He was an enthusiastic observer who recorded what he saw in great detail. The spirit of this youthful work may be judged from an entry in his diary for 1826:

Caught a sea-mouse, Aphrodita Aculeata of Linnaeus; length about three or four inches; when its mouth was touched it tried to coil itself in a ball but was very inert; Turton states it has only two feelers. Does not Linnaeus say 4? I thought I perceived them. Found also 3 Palleta Vulgaris and Solen Siliquor.

There are many other entries of this kind. Clearly, he was ambitious to make *discoveries*—was Turton right, or Linnaeus?—and in the judgement of some of his elders he succeeded. In 1827 he appeared before the Plinian Society, an organization devoted to natural history, and its minutes for 27 March record that:

Mr Darwin communicated to the Society two discoveries which he had made. 1. That the ova of the Flustra possess organs of motion. 2. That the small black globular body hitherto mistaken for the young Fucus lorius is in reality the ovum of the Pontobdella muricata.

At this time Darwin was 18 years old. This may not be evidence of 'genius,' but I venture that most families would find such a son remarkable. In some American universities tenure has been awarded for less.

Perhaps Robert Darwin's attitude was influenced by the fact that in those days natural history was so often carried on by amateurs. He might not have considered it a substantial enough pursuit to be chosen by a gentleman for his life's work—it was more the sort of thing to be pursued as a diverting and scholarly hobby. At any rate, Charles was sent at age 16 to study at Edinburgh, in the hope that he would become a physician like his father. This did not work out well, partly because Charles was too squeamish to bear the sight of operations performed without anaesthetic, which was yet to become available. Moreover, as he later confessed, Charles realized that he would never want for money, and this was 'sufficient to check any strenuous effort to learn medicine'. So, to his father's disappointment, he left Edinburgh after two terms.

Lacking other alternatives, and still thinking his son an unpromising youth, Robert Darwin now proposed that Charles enter the Church. After giving the matter some thought, the boy agreed. A quiet life as a country parson would leave him ample time to pursue natural history; and moreover, as he was later to reflect, there was nothing in his beliefs to rule out such a vocation: 'I did not then in the least doubt the strict and literal truth of every word in the Bible.'

So he went to Cambridge in 1827, and three-and-a-half years later took a degree, still intending to become a clergyman. Although he graduated tenth in his class, he was not known as an outstanding student. By his own account, his time 'was sadly wasted there and worse than wasted':

From my passion for shooting and for hunting and, when this failed, for riding across country I got into a sporting set, including some dissipated low-minded young men. We used often to dine together in the evening, though these dinners often included men of a higher stamp, and we sometimes drank too much, with jolly singing and playing at cards afterwards.

Perhaps this sort of behaviour also contributed to Robert Darwin's despair about his son's future. Recalling these times in his *Autobiography*, however, Charles himself had mixed feelings: 'I know that I ought to feel ashamed of days and evenings thus spent, but as some of my friends were very pleasant and we were all in the highest spirits, I cannot help looking back to these times with much pleasure.'

But Darwin's days at Cambridge were not entirely misspent. He continued to pursue natural history, and became friends with two professors who encouraged him—Adam Sedgwick, the professor of geology, and John Henslow, who according to Darwin 'knew every branch of science'. Both took him on long walking tours studying the countryside. 'I was called by some of the dons "the man who walks with Henslow" ', Darwin said. Henslow's friendship turned out to be especially important, for it eventually resulted in Darwin's invitation to join the crew of HMS *Beagle* as naturalist.

The invitation came just after Darwin had returned home to Shrewsbury. Having completed his formal education, and without any specific prospects, it was not clear what he would do next. A letter from Henslow reported that Captain Robert FitzRoy, who had recently returned from a hydrographic survey of South America, had requested the services of a naturalist for his second voyage, and Darwin had been recommended for the job. The *Beagle* was to circumnavigate the globe, and do a detailed survey of the coasts of Brazil, Argentina, Chile, Peru, and 'some islands in the Pacific'. The journey would last five years.

Unlike two other naturalists, who had already been offered the job and turned it down, Darwin jumped at it. His father at first refused permission, but with the help of an uncle, Charles persuaded him to relent.

The voyage of the *Beagle* was to be the great event in Darwin's life. It would transform him from an aimless, would-be parson into a full-fledged man of science. But why was he chosen for this assignment? It would be nice, but incorrect, to think that he earned the job solely by his reputation as an up-and-coming naturalist. Scientific ability, as it turned out, was only one of the requirements—perhaps it was not even the main requirement—for the job. Captain FitzRoy was looking for a companion. He could not, as Captain, socialize with his men, and naturally he did not want to face five years at sea with no one to talk to. The letter from Henslow put it plainly: 'Capt. F. wants a man (I understand) more as a companion than a mere collector and would not take any one however good a Naturalist who was not recommended to him likewise as a *gentleman*.' So Darwin was taken on to be Captain FitzRoy's dinner-partner, with 'collecting' duties on the side. There was also one other qualification that Darwin possessed: money. The post carried no salary, and indeed, the ship's naturalist would have to pay a large part of his own expenses. It has been estimated that the voyage eventually cost Darwin between £1,500 and £2,000 from his own pocket, a large sum in those days. For comparison, Darwin's lifetime income from all his books—books that sold briskly—was £10,000.

The voyage was costly in other ways as well. For some time Darwin had been in love with Fanny Owen, the daughter of a friend of his father's. When the *Beagle* set sail on 27 December 1831, Charles left hoping that she would wait for him. He was then 22 years old. While in South America, he received word from his sister that Fanny had married someone else.

HOW THE WORLD LOOKED IN 1831

As Darwin set out on his voyage of discovery, what was the state of the science he was to pursue? How did the world look to a young naturalist in 1831? In one way, 1831 was not so long ago. A man born in that year could easily have grandchildren alive today. But when we consider the world of science, 1831 seems part of another age altogether.

It was generally believed that the earth was only a few thousand years old. And why not? Until shortly before Darwin's birth, there was little reason to think otherwise. Nothing that was known about the heavens or the earth required any longer history. Today we know—or at least we think we know—that the universe began about 15 billion years ago in a 'big bang', and that the earth was formed from some of the debris about 4.6 billion years ago. But in 1831 no one could have suspected such an incredible thing. The most widely accepted date for the beginning of creation was 4004 BC, a date which had been calculated from the biblical genealogies by James Ussher, Archbishop of Armagh, and printed in the Authorized Version of the Bible from 1701 on. Some scientists, as we shall see, had already done the work that was to discredit this history. But their views were yet to be generally accepted.

Moreover, it was agreed by most educated people that the earth had been created *pretty much as it is now*—again, there was little reason to think otherwise. As far as the large-scale history of the earth is concerned, things had remained the same for as long as humans had been keeping records: there had always been the same continents, the same oceans, and the same plants and animals inhabiting them. The scriptural account of creation therefore accorded with common experience.

But the scriptural account did more than just furnish a history. It also provided an explanation of *why* things are as they are. The world is full of wonders—plants and animals of the most complicated design, each one exquisitely adapted to its own special place in the natural order. No one was more aware of this than students of nature such as Darwin. But how could this be? How could the world come to contain such wonders? There seemed only two possible explanations. The first, that it all came about by chance, was too far-fetched to accept. The second and only reasonable explanation was that some guiding hand had brought it all about. When Darwin was at Cambridge, William Paley's *Evidences of the Existence and Attributes of the Deity* was required reading for all students. Paley's work, first published in 1802, was the classic presentation of the 'design' argument. In it he declared, 'The marks of *design* are too strong to be gotten over. Design must have had a designer. That designer must have been a person. That person is God.' Only a short time before setting out on the *Beagle*, Darwin had studied this reasoning and had decided that it was irrefutable.

This combination of ideas—that the earth was created by God in the relatively recent past, in pretty much the same state in which we now find it, inhabited by species with permanently fixed natures—is today known as 'creationism'. Creationism now has a bad name. In the hands of Christian fundamentalists, it has degenerated into a dismal pseudo-science, comparable to the shabbiest parapsychology or UFO-ology. But as late as the early nineteenth century, it was still a perfectly reasonable view, accepted by a majority of educated people, and supported, more or less, by such facts as were known. Darwin himself, at age 22, might fairly be described, with some qualifications, as a creationist.

Scientific opinion, however, had begun to change, and by the time the *Beagle* left England, a great shift of thought was underway. At first only a few pioneering scientists were willing to question the old ways of thinking. One of these was James Hutton, a Scottish physician who had taken up geology as a hobby. In his book *Theory of the Earth*, published fourteen years before Darwin's birth, Hutton argued that the prominent features of the earth's surface were produced by such ordinary forces as wind, water, and weather acting slowly and uniformly over a long period of time. (This general approach became known as 'uniformitarianism'.) River canyons are an obvious example. We can see the process of erosion occurring, as natural forces wear away the earth. When we examine the geological evidence of river canyons, they appear to have been formed in just this way. But, Hutton noted, in order for there to have been time enough for this to have happened, the earth would have to be not thousands but millions of years old.

Hutton's argument might have been ignored except for the fact that it helped to explain something else that biologists were worrying about—the fossils. These little rocklike things, which at first had been dismissed as mere curiosities, had structures amazingly similar to the structures of living organisms. But how could rocks mimic living forms? Some biologists suggested that they were the remains of creatures destroyed by Noah's flood—they were the ones that did not make it into the ark. But why should they have turned to stone? The fate of dead organisms is not to turn to stone, but to rot. If Hutton's view was correct, an explanation was possible. As the dead creatures rotted, their natural substance might have been slowly replaced by the stony material in the soil around them. But that would mean that the fossils were the remains of creatures who lived millions of years ago—a disturbing idea, to say the least.

The Industrial Revolution also played a part in our story. In the early nineteenth century, as part of the increasing industrialization, canals were being constructed all over England. In these excavations the strata were clearly exposed, and attracted the attention of many sharp-eyed observers. One particularly keen observer, William Smith—who became known as 'Strata' Smith—noted that each stratum had its own characteristic pattern of rocks *and its own characteristic types of fossil*. To demonstrate the uniform correlation of strata and fossil content, Smith learned to identify different strata by looking at nothing but the associated fossils—shown the fossils, he could tell you from which stratum they had been extracted.

As a result of all this, a new picture of life on earth was emerging. The earth could be seen as millions of years old, and as having been inhabited in the past by creatures now vanished, whose remains are preserved in the fossil record. Moreover, the order in which these creatures lived could be determined by their positions in the strata: those in the lower strata lived earlier than those in the higher strata.

There was more to come. The French anatomist Georges Leopold Cuvier, whose life ended one year after the *Beagle* set sail, made two fundamental contributions. First, he studied the anatomy of different creatures, patiently comparing similarities and differences. (He is therefore credited with founding the science of comparative anatomy.) These studies eventually gave him such knowledge of anatomical relationships that he was able to reconstruct entire animals from only a few bones. This skill was to be invaluable in dealing with the fossils. Often only bits of animals would be found in the strata. Now, based on Cuvier's studies, it would be possible to infer from those bits an intelligent estimate of what the whole animal was like.

Cuvier's second contribution was equally important. Like most biologists of his day, he was intensely interested in systems of biological classification. Drawing on his anatomical studies, Cuvier developed a new and improved scheme, the most sophisticated yet devised. The system was developed, of course, to classify living organisms. But Cuvier discovered that it applied equally well to the fossils. (Thus he is also credited as the founder of palaeontology.) The long-dead animals could be now seen as members of the same families as living animals—as their relatives. Moreover, the way was now open to noticing, systematically and scientifically, the astounding fact that there was an apparent progression in the strata from simpler to more complex forms of life. Roughly speaking, only fossils of invertebrates were found in the lowest strata, while higher up came the remains of fish and sea animals; next birds and reptiles; and finally, in the highest strata, the remains of mammals.

Considering all this, the conclusion fairly leaps out: life has evolved. Species are not immutable. They change. The conclusion leaps out at *us*, because of what we know—everything seems obvious in retrospect—but for people in 1831 things were still not so clear.

If anyone were to accept the possibility of evolutionary change, we might expect it would be Cuvier. But he did not. Although he accepted it as proven that the world is very old, and that different species have lived at different times, he did not think that the different species descended from one another. Instead he hypothesized that the earth has undergone a series of great catastrophes—the last, perhaps, being Noah's flood—in which life had been extinguished, only to be replaced later by new acts of divine creation. This view was known as 'catastrophism', in contrast with uniformitarianism. As odd as it seems today, catastrophism was at one time a widely accepted theory: as T. H. Huxley later observed, in his inimitable style,

A scheme of nature which appeared to be modelled on the likeness of a succession of rubbers of whist, at the end of which the players upset the table and called for a new pack, did not seem to shock anybody.

Knowing all that he did, why didn't Cuvier become a uniformitarian and an evolutionist? Why did he choose instead to indulge in such speculations as ancient 'catastrophes'? The fact that he was a pious man, and wanted an account compatible with the Bible stories, is only part of the story. There is another, deeper reason why evolutionism was resisted within the scientific community. Its basic idea made no sense. Evolution would require that one kind of organism gradually change into a different kind. But how could this possibly happen? What possible mechanism could account for such change? Without a plausible theory of *how* such change could take place, the idea of evolution was too far-fetched to be believed.

The situation may usefully be compared with another, more recent episode in the history of science. For many decades prior to the 1960s, evidence had been accumulating from geology and palaeontology that the continents are not stationary, that they are in motion relative to one another. Indeed, at some time in the distant past it appears that all the southern continents, including India, were bound together in one great land mass covering the South Pole. The continents as we know them today resulted from the break-up of this supercontinent, with its parts gradually drifting apart. Many geological and palaeontological facts made no sense apart from some such assumption as this.

However, very few scientists could bring themselves to believe this. 'Continental drift' made no sense. How could it be possible? Are we to believe that the continents move through the seas like so many giant ships? Or that they slide along the ocean floor like furniture being pushed around someone's living room? In the absence of a mechanism to explain *how* this could happen, the theory of continental drift was too far-fetched to be believed, and respectable scientists did not believe it.

Then in the 1960s an idea was formulated that could explain how continental motion is possible: plate tectonics. The surface of the earth is broken into a few large 'plates' that move relative to one another. The motion is too small to be measured by any but the most sensitive instruments; but it does happen, and the continents ride on these plates. Today, supported by abundant evidence, the theory of plate tectonics is the new orthodoxy, everywhere accepted, and continental drift is no more than a trivial deduction from it.

Continental drift came to be accepted after a process that had four stages:

1. There was the time when no one had ever thought of it.
2. There was the time when growing evidence suggested that it had occurred, but it seemed like a crazy idea because no one could imagine how it was possible—no one could think of a plausible mechanism to explain it—and so scientists rejected the idea, looking instead for other ways to account for the evidence.
3. Then a plausible mechanism was discovered;
4. And as additional evidence was gathered, the idea that previously seemed impossible became widely accepted.

As the *Beagle* put to sea, the situation with regard to evolution had reached the second of these stages. There was much evidence suggesting that evolution had occurred, but no known mechanism could explain how it was possible. So scientists by and large rejected the idea, and tried to find other ways to account for the evidence—by such theories as catastrophism, for example.

However, a few adventurous spirits had chosen to embrace evolutionism anyway. Charles Darwin's grandfather, Erasmus Darwin, had been one of them. In 1794–6 Erasmus Darwin had published a two-volume, 1,400-page work, *Zoonomia, or the Laws of Organic Life*, which included a defence of the idea. The *Zoonomia* had had little impact, because it lacked a coherent account of how evolutionary change might take place. As a work of real science, there wasn't much to it, 'the proportion of speculation', as Charles was later to remark, 'being so large to the facts given'. Charles had read and admired it, but was not convinced.

A more interesting and important effort to support evolutionism was provided by the French naturalist Jean Baptiste Lamarck. Lamarck fully appreciated the need to supply a mechanism for evolutionary change, and tried to do so in his book *Zoological Philosophy*, published the year Darwin was born. Lamarck argued that within every organism there is a force propelling it towards greater complexity and perfection. The unhindered operation of this force would lead naturally to development along an 'upward path', but progress is diverted by environmental pressures—as they are moving along the preordained upward path, organisms must also develop characteristics that allow them to survive in their specific environments. This they do by means of a process called 'the inheritance of acquired characteristics'. It is this subordinate element of his theory that became known to posterity as 'Lamarckism'.

To explain adaptation to local environments (as opposed to the main lines of development dictated by the organism's inner force) Lamarck speculated that the organs of individual animals can be modified through use or disuse as they respond to external conditions. These modifications may then be passed on to their offspring. For an example he chose the giraffe, a recently discovered animal which was the object of much curiosity throughout Europe. Lamarck imagined that the modern giraffe was the descendant of ancient antelopes who ate the leaves of trees. As the antelopes reached after higher and higher leaves, their necks, tongues, and legs would be stretched and would grow a tiny bit longer. Their offspring would then inherit the slightly elongated parts, and the process would be repeated generation after generation.

It was a worthy try, and if it had worked, Lamarck, not Darwin, would have been the father of evolutionary biology. But it didn't work. The idea of an 'internal force', propelling organisms to greater complexity and perfection, never gained wide support, and indeed there was little to be said for it. It was dismissed as mere speculation. On the other hand, the idea that acquired characteristics may be inherited had obvious merit. It rested, first, on the patently correct observation that the bodies of most organisms are plastic, and can change with use and disuse, and second, on the idea that the material that is passed on to offspring ('the germ cells', as they were called in the nineteenth century) interacts with and can be affected by the rest of the body ('the somatic cells'). Thus, the 'inheritance of acquired characteristics' was not at all an unreasonable hypothesis. It was not accepted in Lamarck's lifetime, though, because the hypothesis did not seem to fit the facts. An animal that is naturally skinny, but that becomes muscular through exercise, does not then have muscular offspring. A dog whose leg becomes withered through disuse (perhaps because the leg is bound up for a long time) does not then produce pups with withered legs. Moreover, even if acquired characteristics could be passed on, all the mysteries of adaptation would not thereby be solved. What of the protective coloration that serves so many animals, including the giraffe, as camouflage against predators? Are we to imagine the ancient antelope straining to alter its skin-colour, just as it strained to reach the high leaves?

Despite these difficulties, Lamarck's view was to enjoy a vogue many years after his death. Darwin himself was to accept it and make it a minor part of his own theory. After the fact of evolution was accepted, in the late nineteenth century, Lamarck's reputation grew, and this part of his theory seemed to many scientists a reasonable alternative to Darwin's view. The inheritance of acquired characteristics was not finally set aside until well into the twentieth century.

In any case, Lamarck was certainly on the right track. He was trying to do what needed to be done if evolution was to be proved: he took it as his project to explain how one kind of animal can be transformed, ever so gradually, into another. But, because his explanation was not widely accepted, he died a neglected figure, during Darwin's second year at Cambridge.

An account such as this, explaining the progress of science up to Darwin's time, is inevitably misleading, because it gives the impression of a clear and inexorable march towards truth. But such coherent narratives are the products of hindsight, which enables us to select from the confusion of historical detail just those elements that make up the story we want to tell. For those living through the times, things are never so simple. What did all this look like to Darwin? He was certainly aware of these developments. At Cambridge he had encountered the great geologist Charles Lyell, who was to become a lifelong friend. Lyell's three-volume *Principles of Geology* provided such powerful support for uniformitarianism, and a view of the earth as millions of years old, that catastrophism would soon become a dead theory. The first volume appeared in 1830; Darwin took it with him on the *Beagle*, studied it, and declared it 'wonderfully superior' to any other work he had read. Darwin therefore had come to believe that the world was much older than Bishop Ussher's calculation, and doubtless he was fascinated, as were all students of nature, by the fossils.

But compared to 'the great question of species', these were small matters. The early champions of evolution, no matter how interesting we find them in retrospect, were not the important figures of the day. Darwin had heard of Lamarck, but Lamarck was not a thinker with whom one had to contend, and Darwin was uninterested in Lamarck's project. In the *Autobiography* Darwin recalled an experience at Edinburgh: 'One day, when we were walking together, [a Dr Grant] burst forth in high admiration of Lamarck and his views on evolution. I listened in silent astonishment, and as far as I can judge, without any effect on my mind.' Lyell, who did have an effect on Darwin's mind, refused to accept the mutability of species, and, as the *Beagle* set off, the young naturalist shared his scepticism.

THE VOYAGE OF THE *BEAGLE*

In addition to completing the coastal survey of South America begun on her earlier voyage, the *Beagle*'s mission was to carry out a series of chronological measurements around the world. (The study of natural history was not one of the expedition's primary purposes.) The chronological measurements were necessary for fixing longitudes more precisely. Captain FitzRoy, an expert in the use of such devices, had 22 chronometers on board, and the ship's crew included an instrument-maker to look after them.

In many ways Robert FitzRoy fitted perfectly the stereotype of a nineteenth-century sea-captain. He was an iron disciplinarian, a commanding figure feared and respected by his men. Four months into the voyage, Darwin wrote to his sister: 'I never before came across a man whom I could fancy being a Napoleon or a Nelson. I should not call him clever, yet I feel convinced nothing is too great or too high for him. His ascendancy over everybody is quite curious.' It is even more curious considering that, when the voyage began, FitzRoy was only 26 years old—just four years older than Darwin.

Such men often combine large-scale virtues with equally large-scale vices. So it was with FitzRoy. Darwin wrote of him:

FitzRoy's character was a singular one, with many very noble features: he was devoted to his duty, generous to a fault, bold, determined, indomitably energetic, and an ardent friend to all under his sway. He would undertake any sort of trouble to assist those whom he thought deserved assistance. He was a handsome man, strikingly like a gentleman, with highly courteous manners . . .

Despite these good qualities, however, FitzRoy was not a pleasant companion, and living at close quarters with him for five years was not a happy experience. He was an intolerant, dogmatic man who found it difficult to accept disagreement even from his peers. Darwin complained that conversations with him consisted mainly in FitzRoy talking and Darwin listening. What was worse, he was a Bible-thumping religious fanatic, and a zealous defender of slavery, which Darwin detested. Years later Darwin was to recall that

We had several quarrels; for when out of temper he was utterly unreasonable. For instance, early in the voyage at Bahai in Brazil he defended and praised slavery, which I abominated, and told me that he had just visited a great slave-owner, who had called up many of his slaves and asked them whether they were happy, and whether they wished to be free, and all answered 'No.' I then asked him, perhaps with a sneer, whether he thought that the answers of slaves in the presence of their master was worth anything. This made him excessively angry, and he said that as I doubted his word, we could not live any longer together.

A few hours later, FitzRoy's temper had subsided, and things were back to normal. There can be little doubt that, as much as he might have admired him in some ways, Darwin did not like FitzRoy. In 1839, three years after the voyage was over, Darwin described the captain as 'a man who has the most consummate skill in looking at everything and everybody in a perverted manner'.

Darwin got on much better with the other members of the ship's company. Although he was undoubtedly a 'gentleman', he was no snob, and throughout his life Darwin made friends easily with all manner of people—a quality that helped him in collecting data for his researches from farmers and pigeon-breeders as well as from academics. On board the *Beagle* he was constantly seasick. This, combined with his obvious lack of seafaring experience, could easily have branded him unfit, but he earned the respect of the sailors by his persistence and devotion to his tasks in the face of continual discomfort. Eventually they would volunteer to assist him in various ways.

The one exception was Robert McCormick, the ship's physician. The physician was also slated to work part-time as a naturalist, an arrangement common in nineteenth-century expeditions. In fact, it was McCormick, not Darwin, who was to have been the expedition's 'official' collector. Darwin was present chiefly as the captain's companion. But Darwin had advantages that McCormick could not match: when in port, Darwin had the resources to make lengthy excursions inland while McCormick had to stay aboard ship. Darwin, unlike McCormick, could even hire assistants. Soon Darwin's work was overshadowing McCormick's, and finally, after two years, McCormick left the *Beagle* and went home to England.

Darwin's task was to make close observations of the geology and the flora and fauna of the various places the *Beagle* would visit. This he did with the greatest attention to detail, recording his observations daily in his journals. Five years of this might, for most of us, seem tedious. But for the boy who had been mad for beetles, it was pure joy. Two months into the voyage, the expedition arrived in Brazil, and Darwin wrote:

The day has passed delightfully. Delight itself, however, is a weak term to express the feelings of a naturalist who, for the first time, has wandered by himself in a Brazilian forest. The elegance of the grasses, the novelty of the parasitical plants, the beauty of the flowers, the glossy green of the foliage, but above all the general luxuriance of the vegetation, filled me with admiration . . . To a person fond of natural history, such a day as this brings with it a deeper pleasure than he can ever hope to experience again.

This glorious day came at about the same time as the captain's tirade about slavery. From the outset, then, the voyage was an unpredictable mix of scientific wonder and personal stress.

The expedition spent over three years in and around South America. During this time Darwin made numerous treks inland, at times being separated from his companions for weeks. His itinerary was set largely by the luck of the moment. A typical adventure started when he met 'an Englishman who was going to visit his estate, situated, rather more than a hundred miles from the capital, to the northward of Cape Frio. I gladly accepted his kind offer of allowing me to accompany him.' And so he was off on a three-month jaunt, while the ship was elsewhere making chronological measurements. But it was not always safe to be a stranger in lands torn, even then, by political intrigue. Of one foray into Argentina, Darwin wrote:

The minute I landed I was almost a prisoner, for the city is closely blockaded by a furious cut throat set of rebels. By riding about (at ruinous expense) amongst the different generals I at last obtained leave to go on foot without passport into the City. I was thus obliged to leave my Peon and luggage behind; but I may thank kind Providence that I am here with an entire throat.

Darwin's adventurousness in South America contrasts strikingly with the quiet life he would lead once back in England, where the sixteen miles between Down House and London would often be more than he could manage.

In addition to recording his observations, Darwin collected large numbers of specimens—of fossils, plants, and animals—and sent them back to England, along with many letters describing his work. In the relatively small circle of British scientists, news of this quickly spread, and Darwin began to acquire a reputation. Adam Sedgwick, the Cambridge geologist who had taken an interest in Darwin while he was a student, declared that the young naturalist was 'doing admirable work in South America, and has already sent home a collection above all price ... There was some risk of his turning out an idle man, but his character will be now fixed, and if God spares his life he will have a great name among the naturalists of Europe.'

If Darwin had been only a collector and recorder, he would never have accomplished so much of importance. He did, to be sure, have a great passion for the individual fact. But he was also constantly asking himself questions of a more theoretical kind. His journals are full of musings about such matters as why animals are distributed as they are, and about why some species, whose bones he was gathering, had become extinct. In this frame of mind, and as years of observations accumulated, it would be surprising if a naturalist did not sometimes wonder, however idly, whether the hypothesis of 'mutability' could not help answer some of the puzzling questions. We do not know that Darwin ever put this to himself directly. But we do know that he was constantly reflecting on such questions, and that by the end of the voyage he was in the process of rejecting his old belief in immutability.

In September 1835, having finally left South America, the *Beagle* arrived in the Galapagos Islands, about 650 miles west of Ecuador. This group, consisting of thirteen large islands and numerous smaller ones, was inhabited mainly by Ecuadorian blacks who had been banished for political crimes. The expedition remained in the islands for five weeks; Darwin spent three weeks ashore.

A great deal of lore has grown up around Darwin's visit to these islands, and it is now difficult to separate fact from fancy. We know that, at some time during the five-year voyage, Darwin began to doubt the immutability of species. But when? Was it such a gradual change of mind that no definite date can be assigned? Or were there specific observations that startled him into doubt? In later years Darwin himself encouraged the belief that the weeks in the Galapagos were crucial. But this may have been an unreliable memory, prompted by his later realization that what he observed there illustrates natural selection so perfectly. An examination of his journals does not suggest that, at the time, Darwin had any special reaction to what he saw there.

At any rate, the Galapagos Islands were virtually a laboratory experiment in evolution. They were, geologically speaking, of recent volcanic origin—only a few million years old—so that there would be no 'native' life-forms. The plants and animals there would have spread originally from South America. The islands were close enough together that these original immigrants would have settled on several islands. But they were far enough apart that, once settled, the plants and animals would from separate population-groups that would breed independently of one another. Finally, enough time had passed for there to have been significant adaptation to local conditions. Thus all the conditions were right for evolutionary change to be readily observable. The inhabitants of the islands would have evolved, from a common parent stock, into different species, as they adapted to local circumstances. Looking back, Darwin later realized that this was exactly what he had observed.

The popular version of the story is that Darwin was especially impressed with the tortoises and the finches found on the islands. In his journals he gave a lengthy description of the tortoises, adding with boyish exuberance that 'I frequently got on their backs, and then giving a few raps on the hinder part of their shells, they would rise up and walk away;—but I found it very difficult to keep my balance.' Most of his observations on the tortoises are unremarkable; they are similar to the kinds of things he says about all the species he describes. But then he adds this:

I have not as yet noticed by far the most remarkable feature in the natural history of this archipelago; it is, that the different islands to a considerable extent are inhabited by a different set of beings. My attention was first called to this fact by the Vice-Governor, Mr Lawson, declaring that the tortoises differed from the several islands, and that he could with certainty tell from which island any one was brought. I did not for some time pay sufficient attention to this statement, and I had already partially mingled together the collections from two of the islands. I never dreamed that islands, about fifty or sixty miles apart, and most of them in sight of each other, formed of precisely the same rocks, placed under a quite similar climate, rising to a nearly equal height, would have been differently tenanted . . . I obtained sufficient material to establish this most remarkable fact in the distribution of organic beings.

'This most remarkable fact' would turn out to be immensely important later on when Darwin would finally solve the mystery of how evolution takes place.

The finches were even more remarkable. Darwin found thirteen separate species of finches, each inhabiting a different island, and each adapted to its own specific environment. In particular, each had a differently shaped beak, adapted to different types of food. On one island nuts were plentiful, and the finches of that island had beaks suitable for cracking and eating nuts; on another island the available food was insects, and the finches had beaks good for grabbing insects; and so on. After describing the variety of beaks, Darwin wrote:

Seeing this gradation and diversity of structure in one small, intimately related group of birds, one might really fancy that from an original paucity of birds in this archipelago, one species had been taken and modified for different ends.

These birds are today called by zoologists 'Darwin's finches'.

When we look at these facts, and Darwin's comments on them, it seems obvious that Darwin must have been on the verge of abandoning his belief in the immutability of species. 'Descent with modification' was staring him in the face: the tortoises and the finches were descendants of common ancestors, 'modified', as Darwin says, 'for different ends'. And so the popular story is that, confronting the inhabitants of the Galapagos, Darwin became an evolutionist.

It's a good story, but unfortunately it probably is not true. As for the tortoises, they were not, in fact, separate species, even though they differed from island to island. They were mere variations of one species. The fact that a species with an extended range develops variations was well-known to all naturalists, including Darwin. (It is the unremarkable fact that humans from Africa are recognizably different from humans from Scandinavia.) Once it became clear that the tortoises were only variations, Mr Lawson's 'revelation' that they were recognizably different should have been nothing exciting.

As for the finches, they *were* separate species, and so they are much more interesting; and the famous sentence quoted above certainly does seem to be an undeniable recognition that 'descent with modification' had taken place. But that sentence was added to Darwin's account when a new edition of his journals was published in 1845—well after the theory of natural selection had been formulated. It does not appear in earlier editions. Moreover, the finches do not appear at all in *The Origin of Species*, even though they perfectly illustrate the theory of natural selection, and that book is full of comparable examples. So it seems unlikely that they occupied the prominent place in Darwin's thoughts that legend suggests.

The upshot is that, although the visit to the Galapagos was, along with many other episodes on the voyage of the *Beagle*, important in undermining Darwin's faith in immutability, it was not critically important. It is more likely that Darwin was gradually converted to evolutionism, not by one or two startling discoveries, but by the combined influence of many observations, including these. Perhaps, in retrospect, what he saw on these islands seemed especially important because the stop there came relatively late in the voyage, at a time when the accumulation of other experiences, and other questionings, had already begun to work on his mind.

Of course, modern readers are mainly interested in the *Beagle* voyage because of its part in propelling Darwin towards the theory of natural selection. This is hindsight. However, the voyage had another result, having nothing to do with evolution, that was of great immediate interest to scientists of the time, and that secured Darwin's reputation as a rising man of science. While on the voyage Darwin solved a mystery that had long puzzled geologists: the formation of coral reefs. Atolls, 'those singular rings of coral-land which rise abruptly out of the unfathomable ocean', were utterly inexplicable. Why should corals form circles? And how can they form these circles 'out of the unfathomable ocean', when we know that corals cannot live at depths of more than 20 to 30 fathoms? On what are the atolls based?

Darwin's explanation was elegant and simple. He proposed a theory according to which corals grow in the shallow water surrounding an island. Originally, the coral growth snuggles against the island. But then geological forces cause the land to begin sinking slowly beneath the sea. The living coral, however, continues to grow upward, following its original circular pattern, until all that remains is the coral ring, outlining the shape of the now-vanished island. A similar explanation also accounts for the coral reefs that surround islands and continents: in these cases, the 'subsidence' of the land mass has begun but has not yet progressed so far as in the atoll.

The pattern of Darwin's explanation has some interest. In his autobiography he notes that 'the whole theory was thought out on the west coast of South America before I had seen a true coral reef'. It was, then, in a sense, speculation. But it was the kind of inspired guesswork that is indispensable in science. Darwin had asked himself, given the facts as known, what could possibly account for them? Then, as the *Beagle* sailed west across the Pacific, Darwin was able to confirm and verify his explanation by detailed examination of the reefs and atolls they would encounter. The theory of natural selection was to be formulated in a similar spirit. Known facts create a puzzle; a theory is proposed that, if confirmed, could solve it; and then further investigations confirm the theory.

After visiting Australia, and stopping at several islands on the way home, and then returning briefly to South America, the *Beagle* completed its circumnavigation of the globe. The ship's company arrived back in England on 2 October 1836. Darwin later reflected, 'I have always felt that I owe to the voyage the first real training or education of my mind', and added:

As far as I can judge of myself I worked to the utmost during the voyage from the mere pleasure of investigation, and from my strong desire to add a few facts to the great mass of facts in natural science. But I was also ambitious to take a fair place among scientific men—whether more ambitious or less so than most of my fellow-workers I can form no opinion.

Gone were any thoughts of a country parsonage; Darwin's sole ambition was now to 'take his fair place' among men of science.

Darwin's long book about the trip, today known as *The Voyage of the Beagle*, was published three years later and was an instant popular success. To the untravelled Englishmen of the 1840s, it was a fascinating saga of exotic lands. Along with the geology and natural history, Darwin included tales of pirates' skulls, dangerous rebels, fantastic beasts, cannibalism, and murder. Another reason for the book's success was Darwin's likeable personality. Written in the first person, the book is full of Darwin the man, and his character shines through—ebullient, humane, witty. The book even had a distinct moral tone: he inveighs against slavery on more than one page, and his observations are among the most moving in abolitionist literature. Here is what he says about one encounter with a slave in Brazil:

I may mention one very trifling anecdote, which at the time struck me more forcibly than any story of cruelty. I was crossing a ferry with a negro who was uncommonly stupid. In endeavoring to make him understand, I talked loud, and made signs, in doing which I passed my hand near his face. He, I suppose, thought I was in a passion, and was going to strike him; for instantly, with a frightened look and half-shut eyes, he dropped his hands. I shall never forget my feelings of surprise, disgust, and shame, at seeing a great powerful man afraid even to ward off a blow, directed, as he thought, at his face. This man had been trained to a degradation lower than the slavery of the most helpless animal.

Darwin wrote three other books, more strictly scientific, about his geological investigations on the voyage. His 'fair place among scientific men' was swiftly secured.

As for Captain FitzRoy, he also wrote a book about the voyage, claiming that it had produced evidence confirming the truth of the creation story in Genesis. He would become more and more a resentful and tragic figure. He was soon jealous of Darwin's success, and as Darwin's fame grew, his bitterness increased. His own career could hardly be termed a failure. In the 1840s he was, for a time, made governor of New Zealand. But his religious views hardened, and he saw Darwin as a blot on his personal record. When Darwin finally published *The Origin of Species* in 1859, and began by citing the importance of the voyage in shaping his thought, it was too much for FitzRoy, who came to blame himself for Darwin's heresy.

In blaming himself, FitzRoy may have been correct, although not in any way he could have recognized. Five years spent with such a man, while in one's twenties, is bound to leave its mark. Stephen Jay Gould's speculation seems eminently reasonable:

And think of Darwin's position on board—dining every day for five years with an authoritarian captain whom he could not rebuke, whose politics and bearing stood against all his beliefs, and whom, basically, he did not like. Who knows what 'silent alchemy' might have worked upon Darwin's brain during five years of insistent harangue. Fitzroy may well have been far more important than finches, at least for inspiring the materialistic and anti-theistic tone of Darwin's philosophy and evolutionary theory.

Six years after publication of the *Origin*, and after publicly denouncing Darwinism on more than one occasion, Captain FitzRoy, now a pathetic figure, committed suicide.

THE DISCOVERY OF NATURAL SELECTION

Upon returning to England, Darwin took up residence in London. He was soon accepted into the Geological Society, a sign that he had 'arrived' scientifically. He would frequently speak before the Society—his recent travels gave him firsthand knowledge of many subjects its members were interested in—and within a year he was made an officer of the group. His entry into this exclusive club was facilitated by Sedgwick, who, without Darwin's knowledge, had been reading Darwin's letters aloud at its meetings as they arrived from South America. Lyell, the leading figure in this circle, now became one of his closest friends. It was now plain that Darwin would not be an 'idle man', or a parson. He would devote full time to his scientific pursuits.

In July 1837, after ten months at home, Darwin opened his first 'Transmutation Notebook'. This was his first unequivocal declaration (at least, the first of which we are aware) that he was now an evolutionist. He never used the term 'evolution'; he preferred, at first, 'transmutation', and later, 'descent with modification'. But whatever terms were used, the notebooks were to record his investigation into the 'mystery of mysteries', the question of species. He now accepted that species evolve, and he set himself to discover how. Like Lamarck before him, he would try to identify the mechanism of change.

For the next two years, Darwin led two lives. Publicly, he was involved in research that had nothing to do with the question of species. He addressed the Geological Society on such topics as the role of earthworms in soil formation, and the geological origins of the 'parallel roads' of Glen Roy in the Scottish Highlands. He worked on his book about the voyage of the *Beagle*. He attended to the disposition of the collections he had brought back from the trip. All the while, privately in his notebooks, he sought the answer to Lamarck's question, a project which preoccupied him more and more.

By the end of 1839, he had discovered the key to solving the riddle. He had formulated a theory, beautiful in its simplicity, which he would later call 'natural selection'. The theory sees evolutionary change as the inevitable consequence of three obvious facts:

The geometrical increase of populations. If left unchecked, the size of the population of a single animal or plant species would increase until the world is overrun. One parent might produce 10 offspring; each of those might then produce a similar number; and so on. In the first generation there will be 1; in the second generation, 10; in the third, 100; in the fourth, 1,000; and so on *ad infinitum*.

Variation. Individuals within the same species are not always exactly alike—they differ in their particular characteristics.

Inheritance. Individuals tend to pass on their own particular characteristics to their descendants.

What happens when we look at these three facts together? First we notice that, obviously, not all the animals (or plants) can survive. If they did, the earth would quickly be overrun. There will, therefore, be a competition for survival, in which some will live and some will die. Now if there are variations among the animals—differences between them—this means that some individuals will have an advantage in the struggle; for at least some of their differences will inevitably *make a difference* in the individuals' abilities to secure food, avoid predators, and so on. The individuals with the advantageous characteristics will be more likely to survive and reproduce, and so will be more likely to pass on their characteristics to the next generation. The individuals that did not survive will not be able to pass on their particular characteristics. Therefore, future generations will tend to resemble the individuals who have the advantages. In this way, the characteristics of species will change, and when enough of their characteristics have changed, there will be a new species. There is more to the theory, but we will come to that later. This is enough to show the basic idea.

The theory of natural selection was, in its most elementary form, a deduction from simple facts known to everyone. Darwin did not discover the facts. His genius was to consider them together, to recognize the pattern they formed, and to notice their implications. This is not an uncommon phenomenon in science. In the first few years of the twentieth century Albert Einstein was to make a similar discovery. Like the theory of natural selection, the theory of special relativity was also an unexpected deduction from facts already known to all scientists.

The period between 1837 and 1839 has been subjected to intense scrutiny by scholars wanting to know how Darwin happened upon his theory. What process of reasoning led him to it, when others had considered the question and had come up empty? In his *Autobiography*, Darwin himself provided one answer. He credited reading Thomas Malthus's book *An Essay on the Principle of Population* with triggering the discovery:

In October 1838, that is, fifteen months after I had begun my systematic inquiry, I happened to read for amusement Malthus on Population, and being well prepared to appreciate the struggle for existence which everywhere goes on from long-continued observation of the habits of animals and plants, it at once struck me that under these circumstances favorable variations would tend to be preserved, and unfavorable ones to be destroyed. The result of this would be the formation of new species. Here, then, I had at last got a theory by which to work . . .

If this is taken at face value, it would seem that Darwin thought of natural selection all at once, as a single blinding insight. But like the story of the finches, this one also turns out to be too good to be true. The *Autobiography* was written almost forty years later; it records an old man's fallible memory. Better evidence is provided by his notebooks, with daily entries made at the time Darwin was struggling to formulate his theory. Those notebooks record no such blinding insight. When Darwin read Malthus, he did not rush to set down a great discovery. He only noted another interesting observation which might eventually fit into the over-all picture. The truth seems to be that the theory was formulated, after many false starts, slowly, as the result of many suggestive observations, of which Malthus provided only one. The notebooks show him trying out first one theory and then another; asking a variety of questions, and recording various possible answers; and generally juggling a multitude of ideas until he pieced everything together.

In 1839, as the theory was finally being worked out, Darwin married his cousin Emma Wedgwood. Emma was the daughter of the same uncle who had helped persuade Robert Darwin to allow Charles to go with the *Beagle*. A pious woman, Emma would never accept her husband's theories. She knew that she was marrying an unorthodox man. Shortly after their marriage, she wrote Charles a moving letter, professing her love, but also describing the pain his free-thinking caused her. He would never change her mind, nor she his. Thus Darwin joined the ranks of famous thinkers who could not convince their own spouses. But it was a happy marriage, and like a good upper-class wife Emma would take it as her duty to provide a tranquil environment for her husband's work. They lived in London for three years, and then in 1842 moved to Down House in a small village sixteen miles from the city, where they would remain for the rest of their lives.

THE LONG DELAY

Darwin was welcomed into the little village of Downe as a celebrity. By this time, his book on the voyage of the *Beagle* was well known. In time he would become the village's leading citizen, the confidant of the local parson, and a faithful contributor to local causes. In old age he would become the town magistrate, the arbiter of disputes between his neigh-bours. His completely conventional, respectable life would serve him well when he became the centre of controversy. No matter how scan-dalous his ideas, no matter how threatening they might seem to religion and morality, Darwin personally would be above reproach.

At Down House he and Emma raised eight children; two more died in infancy. It was a model family. The one cloud in Darwin's personal life was his health. For three decades he suffered from chronic weakness. Often he complained that he could not work for more than an hour without lying down. Often he would not go into London, saying he did not feel up to it. We do not know exactly what was wrong with him. For a long time it was thought he suffered from Chagas's disease, a malady related to African sleeping sickness, which he could have contracted in Argentina. Another possibility is brucellosis, or undulant fever, also common in Argentina. Some have suggested that Darwin's problems might have been psychological in origin, related perhaps to his anxiety over the nature of the theory he was developing and the scandal it was sure to provoke. Whatever its cause, bad health plagued him for most of the rest of his life, and constantly interfered with his work.

At about the time he moved to Downe, Darwin wrote a 48-page 'sketch' of the theory of natural selection. Then in 1844 he produced a long Essay, of 230 pages, presenting the theory in greater detail, and made arrangements to ensure that it would be published if anything should happen to him.

It would seem that Darwin was now ready to announce his theory. He had it all worked out, and had amassed a great deal of evidence in its support. He had written an elegant formal presentation. The Essay of 1844 was no mere outline. Although it lacked the wealth of detail that would grace the *Origin*, it was very much a finished book, and it is only a slight exaggeration to say that it was as well crafted as anything he ever wrote. But he did not publish it. He did not even continue to work on it. Instead, he set it aside and turned to an eight-year study of barnacles— conventional, unexciting work that could have been done by any other naturalist. He would not publish anything on the species question for another fourteen years, and even then he would publish only because he was forced to do so.

Why did Darwin delay for so long? At the very moment when he had solved the greatest problem in natural history, why did he turn away from it? It is one of the fascinating questions in the history of science. Darwin himself provided part of the answer when he admitted that he was reluctant to publish a work that would cause distress to his friends and family. It was obvious that his theory would be taken as an attack upon Christianity, even if he himself avoided all reference to religion. In the furious controversy that would surely follow, his name would be identified with atheism and worse. Darwin, who was proud of his growing reputation as a sober scientist, did not relish such controversy for himself, and even less for his pious wife.

Thus Darwin would welcome excuses not to publish, and it was easy to find them. He began to tell his friends that the theory should not be made public until he had got everything perfect. He wanted, he said, to present such overwhelming evidence that even the most sceptical scientists would have to take the theory seriously. If his reasoning contained the slightest flaw, the whole idea was likely to be dismissed. Moreover, he knew there would be countless objections to the theory, and he wanted to anticipate and answer as many of them as possible in advance. All this, he thought, must be done to 'secure' the theory, and it would take a really big book—many times the length of the Essay of 1844.

But why, instead of working on that big book, did he take up the study of barnacles, a topic apparently unrelated to the species question? Despite the success he had already achieved, Darwin felt that his training as a scientist was incomplete. Much of his work until now had been in geology, and in biology his work had been largely superficial. Consider, for example, the famous finches. He did not, on his own, even recognize them all as finches. An ornithologist at the British Museum had to point this out to him. He had never done the kind of painstakingly detailed work by which professionals prove their mettle. Without such experience, he would surely commit errors of detail in his 'big book'— errors that would cause biologists to dismiss it out of hand. Thus he felt he needed to undertake a type of project—the detailed, fully professional investigation of a single species—that he had not, up to this time, attempted. The work on barnacles was to be that project.

The distinguished botanist Joseph Hooker was at this time Darwin's closest friend. He later wrote,

It is impossible to say at what stage of progress [Darwin] realised the necessity of such a training as monographing the Order offered him; but that he did recognize it and act upon it as a training in systematic biological study, morphological, anatomical, geographical, taxonomic and descriptive, is very certain . . .

Hooker reported that, in conversation, Darwin described 'three stages in his career as biologist, the mere collector, in Cambridge etc.; the collector and observer, in the 'Beagle' and for some years after; and the trained naturalist after, and only after, the Cirripede [barnacle] work'. Thus it appears that the barnacle work was not a flight from the species question after all. It was, in Darwin's view, a necessary preliminary to pursuing the big issue.

There is one other piece to the puzzle. Howard Gruber has argued that the traditional explanation of Darwin's delay, which I have just outlined, is insufficient to account for all the facts. Surely, he says, the need to complete his education cannot explain Darwin's spending *eight years* on the barnacles—not while he sat on the most important discovery in the history of his subject. The rest of the explanation, Gruber says, is fear—and not merely fear of the reaction to a defence of evolution. Evolution was, by the 1840s, a familiar enough idea that its shock-value was no longer so great. Instead, Darwin feared an adverse reaction to the materialistic philosophy at the core of his particular form of evolutionism. Other evolutionists were speaking of vital forces, guiding spirits, and the directing power of the mind, but for Darwin there was only random variation and natural selection. In his notebooks Darwin ridiculed the ancient idea of the soul as an immaterial thing separate from the body; thought, he said, is a 'secretion of the brain'. ('Oh you materialist!' he adds.) Thirty years later, he would publicly describe all of man's 'higher qualities' as the product of material forces. But the young Darwin knew better than to do that. He knew that

In virtually every branch of knowledge, repressive methods were used: lectures were proscribed, publication was hampered, professorships were denied, fierce invective and ridicule appeared in the press. Scholars and scientists learned the lesson and responded to the pressures on them. The ones with unpopular ideas sometimes recanted, published anonymously, presented their ideas in weakened forms, or delayed publication for many years.

While Darwin was a student, one of his friends had read a paper with a materialist slant at the Plinian Society in Edinburgh, and afterwards all references to it—including the statement of his intention to read the paper, in the minutes of the previous meeting—were stricken from the records. Darwin got the message, and would not publicly advocate such materialism for many years.

Darwin may have feared to announce his theory; but he must have known that delay also had its dangers: there was the risk that someone else would discover the theory and publish it first. And as Darwin dawdled, evolutionism was more and more in the air. In the 1850s Herbert Spencer, a maverick philosopher, produced a series of popular works advocating what he called 'the Development Hypothesis'. Spencer was no scientist, and he had no positive theory about how evolution might have occurred, but he had a pregnant phrase, 'the survival of the fittest', which Darwin was later to adopt. And Spencer was not the only one trumpeting such ideas. In 1844, just as Darwin was completing his long Essay, a book had appeared called *Vestiges of the Natural History of Creation*, written by Robert Chambers but published anonymously. It was a hodgepodge of ideas, and riddled with scientific errors, but it advocated evolutionism (man, said Chambers, is descended from a frog) and it was a best seller. Darwin read the book and was relieved to find that Chambers had not scooped him. But, with so much public interest in 'the species question', it was only a matter of time until someone would.

Darwin and Wallace

Among the readers of *Vestiges* was Alfred Russel Wallace, a young surveyor-turned-botanist. After reading the book, Wallace, a teacher at

Leicester, became obsessed with the question of species. Like Darwin before him, he accepted that species change, and was convinced that the changes must be governed by simple laws of nature. He vowed to discover those laws.

From this point on, Wallace's story is strikingly parallel to Darwin's. First, Wallace set out on his own voyage of discovery. Unlike Darwin's voyage, however, Wallace's expedition was undertaken specifically for the purpose of gathering evidence relevant to the question of species. He went to the Amazon in 1847 and remained there, collecting specimens, for three years. On the voyage home, however, the ship caught fire and his collections were destroyed. Wallace himself barely escaped with his life. Undaunted, he set out again, for the Malay Archipelago, where he would continue his investigations. While there, he too read Malthus's book on population, and as a result discovered for himself the theory that Darwin had formulated, but had not published, years before. Wallace had no such hesitation. The basic idea could be presented in a brief space, and Wallace saw no reason not to do so. He promptly wrote a short paper and sent it off to England. He sent it, in fact, to Darwin.

Wallace's letter arrived in June 1858. Darwin was thunderstruck. 'If Wallace had my MS sketch written out in 1842,' he moaned, 'he could not have made a better short abstract! Even his terms now stand as heads of my chapters.' Having delayed for almost twenty years, Darwin had finally been scooped.

The story of Wallace's letter is well known; it is one of the most dramatic episodes in the history of science. But the fateful letter was not entirely a bolt from the blue—Darwin had some reason to expect, or at least to fear, that something like this might happen. Wallace had let it be known that he was on the trail of such a theory. In 1855 he had published an article in the *Annals and Magazine of Natural History* with the title 'On the Law which has Regulated the Introduction of New Species'. He did not yet have an account of the 'law', but he was obviously a talented man headed in the right direction, and the appearance of this article made Darwin and his friends nervous. Lyell urged Darwin to delay no longer. Darwin remained obstinate; he still would not publish—but he did begin to assert his claim to the theory. He wrote to Wallace, complimenting him on his paper, but also letting Wallace know that he had already staked out this particular territory: 'I can plainly see', he said, 'that we have thought much alike and to a certain extent have come to similar conclusions . . . This summer will make the 20th year (!) since I opened my first note-book, on the question how and in what way do species differ from each other. I am now preparing my work for publication.' However, Darwin would not divulge the exact nature of his work.

'It really is *impossible*', he continued, 'to explain my views (in the compass of a letter), on the causes and means of variation.'

Having said that it was impossible to explain his theory in the compass of a letter, Darwin then proceeded to do so—but not to Wallace. Five months later he wrote a long letter to the American botanist Asa Gray, describing the idea of natural selection in some detail. There was no obvious reason for Darwin to do this, other than to get his work 'on the record' and establish his priority before a neutral witness. At any rate, three months after the letter to Gray, we find Darwin writing again to Wallace, repeating his claim of priority, and again saying that the theory could not be explained in a mere letter: 'I believe I go much further than you; but it is too long a subject to enter on my speculative notions.' Darwin's anxiety was plain. It could hardly be more obvious that he regarded Wallace as a dangerous rival.

Wallace, for his part, saw Darwin as a friendly and eminent colleague, working on the same problem, who had been kind enough to write him a series of letters. It seems clear, then, that Wallace sent his paper to Darwin because of this previous correspondence. Darwin had told Wallace: 'We have thought much alike.' It was natural enough for Wallace to want to share his new idea with the distinguished naturalist and solicit his opinion.

In mid-1856 Darwin had at last begun the serious work of expanding his Essay of 1844 into a 'big book'. The work went slowly, for once again Darwin was gripped by the notion that every detail must be irrefutably proven, every possible objection answered. He now envisioned a book four times longer than the *Origin* would turn out to be; we can only speculate whether that book would ever have been completed. But the arrival of Wallace's letter, two years later, changed everything.

The paper that came with the letter was entitled 'On the Tendency of Varieties to Depart Indefinitely from the Original Type'. Darwin was at a loss about what to do with it. His first reaction was, characteristically, that he should renounce all claim of priority, and have Wallace's paper published instead. He told Lyell, 'I shall, of course, at once write and offer to send it to any journal. So all my originality, whatever it may amount to, will be smashed.' Lyell, however, did not want to see Wallace steal his friend's thunder, and he proposed a Solomon-like solution: the two should publish simultaneously. He and Hooker would arrange to have Wallace's paper *and* a similar paper by Darwin read at a meeting of the Linnean Society, and then published together in the Society's journal. And, Lyell said, this should be done quickly: so quickly, in fact, that Wallace, still in Malay, would not even know what was happening.

In his heart, Darwin knew that this would not be altogether ethical. Wallace had said nothing about publication; he had merely sent Darwin his paper as part of a friendly communication, because he knew Darwin would be interested. Responding to Lyell's suggestion of a joint presentation, Darwin emphasized his doubts about the proposal, and concluded that he could not shake his original feeling that 'it would be dishonourable in me now to publish'. But, after the hand-wringing was done, Darwin agreed, and Lyell's plan was carried out. On July 1, less than two weeks after the arrival of Wallace's letter, Wallace's paper and a short paper by Darwin were both read to the Linnean Society. Darwin was not there; one of his sons had died of scarlet fever two days before. Darwin's friends, however, made sure that the members of the Linnean Society knew who had priority. In addition to the two papers, they also read Darwin's letter to Asa Gray, and Hooker assured the Society, in writing, that he had seen Darwin's Essay fourteen years previously. When the papers were published, these supplementary documents were published with them.

There was, indeed, no doubt about the matter of priority; nor was there any doubt about the relative ranks of the two naturalists: Wallace was a fine scientist, but Darwin was a great genius. Nevertheless, the whole episode had a bad smell. Darwin was a member of the inner circle of British science, surrounded by friends eager to protect his interests, whereas Wallace was a naïve outsider. That, as much as their respective scientific talents, determined what happened. Imagine how Wallace could have felt upon learning that his paper, sent innocently to Darwin, had been published, without his permission, surrounded by three documents designed to minimize its impact. If he had sent his paper directly to a publisher, for example to the *Annals and Magazine of Natural History*, the story would have been very different. Darwin, who had meekly acquiesced in this shabby business, always had a bad conscience about it. But Wallace was a gracious man, and he readily acknowledged that Darwin's work was not only older but deeper than his own. Darwin's relief was palpable. Upon receiving further correspondence from Wallace, in which Wallace expressed pleasure about the joint publication, Darwin wrote to Hooker: 'I admire extremely the spirit in which they are written. I never felt very sure what he would say. He must be an amiable man.'

In fact, Wallace was delighted with the whole business. His main reaction was to be flattered that Darwin and the others thought so well of his work. Upon learning what had happened, he wrote to his mother:

I have received letters from Mr Darwin and Dr Hooker, two of the most eminent naturalists in England, which has highly gratified me. I sent Mr Darwin an essay on a subject on which he is now writing a great work. He showed it to Dr Hooker and Sir C. Lyell, who thought so highly of it that they immediately read it before the Linnean Society. This assures me the acquaintance and assistance of these eminent men on my return home.

Lyell and Hooker had assured the Linnean Society that Darwin had already written on natural selection at great length. Now Darwin was forced to produce a longer presentation of the theory, and quickly. The following year, in 1859, he published *On the Origin of Species by Means of Natural Selection, or the Preservation of Favoured Races in the Struggle for Life*, a work he termed an 'abstract' of the big book that now would never be completed. Thanks to Wallace, the twenty-year delay was over. Wallace wrote to Darwin in 1864:

As to the theory of Natural Selection itself, I shall always maintain it to be actually yours and yours only. You had worked it out in details I had never thought of, years before I had a ray of light on the subject, and my paper would never have convinced anybody or been noticed as more than an ingenious speculation, whereas your book has revolutionized the study of Natural History, and carried away captive the best men of the present age. All the merit I claim is the having been the means of inducing *you* to write and publish at once.

For his part, Darwin would in later years refer to the theory of natural selection not as 'my view' but as 'Wallace's and my view'.

'ONE LONG ARGUMENT'

The presentation of the papers by Darwin and Wallace at the Linnean Society had fallen flat. No great controversy ensued, and when Thomas Bell, the Society's president, later made his annual report, he said that the year had been uneventful. The publication of the *Origin*, however, was another matter. The book was an instant hit. The publisher sold out his first printing of 1,250 copies in one day, and immediately began work on a second. Darwin was a wonderfully clear writer, and the book was easy for any educated person to read and understand. It may have been the last great work of science that will ever be so accessible to the layman. The public was intrigued, the Church was alarmed, and the scientific community was compelled to take evolutionism more seriously than it ever had before.

Later, in the *Autobiography*, Darwin would remark that 'Some of my critics have said, "Oh, he is a good observer, but has no powers of reasoning." I do not think that this can be true, for the *Origin of Species* is one long argument from beginning to end, and it has convinced not a few able men.' Although the book contains many careful observations of natural phenomena, its core is not so much a series of observations as a chain of reasoning. The argument is beautifully simple, appealing to facts evident to the plainest common sense. (It is so simple that, when T. H. Huxley first saw the argument, his reaction was 'How extremely stupid not to have thought of that!' Huxley, who was to become Darwin's staunchest defender, had previously been sceptical about evolutionary hypotheses.) Moreover, unlike other fundamental scientific theories, his reasoning required no abstruse mathematics—for which Darwin was grateful, for as he often lamented, he possessed no mathematical gifts.

The core argument of the *Origin* can be summarized as follows:

1. Organisms tend to reproduce in such numbers that, if all survived to reproduce again, they would soon overrun the earth. (This is the Malthusian observation that population, if unchecked, increases geometrically.)

2. This does not (and could not) happen. No species can continue to multiply unchecked. Each population reaches a certain maximum size, and then its growth stops.

3. It follows that a high percentage of organisms must die before they are able to reproduce.

4. Therefore, there will be a 'struggle for existence' to determine which individuals live and which die. What determines the outcome of this struggle? What determines which live and which die? There are two possibilities: it could be the result of random causes; or the reason could be related to the differences between particular individuals.

5. Darwin admits that sometimes it is random; that is, the reason one organism survives to reproduce, while another does not, may sometimes be attributable to causes that have nothing to do with their particular characteristics. One may be struck by lightning, while another is not, and this may be mere luck.

6. But, he says, we can see that sometimes it is a matter of differences between individual organisms. Consider:

    (a) There are differences ('variations') between members of species. Darwin did not know how or why such variations occur. But it is evident that they *do* occur.

    (b) Some of these differences will affect the organism's relation to its environment, in ways that are helpful or harmful to its chances for survival.

    (c) Therefore, because of their particular characteristics, some individuals will be more likely to survive (and reproduce) than others.

7. Organisms pass on their characteristics to their descendants. Again, Darwin did not know exactly how this happens, but evidently it does: an organism's offspring tend to have its particular characteristics.

8. Therefore, the characteristics that have 'survival value' are passed on, and tend to be more widely represented in future generations, while other characteristics tend to be eliminated from the species.

9. In this way, a species will be modified—the descendants of the original stock will come to have different characteristics from their forebears.

10. At first we call the 'different' organisms a new variety; but when enough of these modifications have accumulated, we call the result a new species. Varieties, then, are 'incipient species'.

The Analogy with Artificial Selection

Darwin did not begin the *Origin* by laying out this argument. Instead, his strategy was to start with a discussion of the work of breeders, who deliberately produce 'improvements' in plants and animals by selective mating. The work of breeders was familiar to all naturalists, and Darwin realized that, if he could show that the process of natural selection is analogous to the activity of breeders, he would give his theory a kind of instant plausibility.

What do breeders do? Suppose we want sheep to have shorter legs, so that they can be more easily confined in pens. We choose, from among the existing flock, the individuals that have the shortest legs, segregate them, and breed them with one another. Their offspring will have, on the average, shorter legs than the rest of the flock. Then we eliminate the longer-legged individuals from the second generation, and repeat the process. Eventually we will have what we want, a strain of sheep with much shorter legs. Further, suppose that during this process we notice that some of the shorter-legged sheep have slightly richer wool. Finding this desirable, we separate out those individuals, and breed only the shorter-legged, richer-wooled animals. Eventually we have a strain in which both desirable characteristics are dominant. Since all shepherds will want to raise this kind of sheep, this strain will soon be more numerous than any other, and we will come to think of these as 'typical' sheep. (And what if, in time, the origin of these characteristics is forgotten? People might come to marvel that God has provided animals so well adapted to human needs.)

'The key', Darwin says, 'is man's power of accumulative selection: nature gives successive variations; man adds them up in certain directions useful to him. In this sense he may be said to have made for himself useful breeds.' In addition to the sheep, countless other examples could be given. Dogs are 'modified' for show by selectively breeding for longer snouts, or glossier coats. Flowers are bred to match our standards of beauty. Darwin himself mentions strawberries:

No doubt the strawberry had always varied since it was cultivated, but the slightest varieties had been neglected. As soon, however, as gardeners picked out individual plants with slightly larger, earlier, or better fruit, and raised seedlings from them, and again picked out the best seedlings and bred from them, then (with some aid by crossing distinct species) those many admirable varieties of the strawberry were raised which have appeared during the last half-century.

This process has affected the development of many thousands of varieties of plants and animals. Indeed, as Darwin notes, virtually all of our cultivated plants and domesticated animals have been produced or modified by such a process. That is why they serve our needs so well, and it is why they are so different from wild varieties. The process may be unconscious as well as conscious: plants and animals that have characteristics important to us may be cherished and protected, while others, of no interest to us, are allowed to perish—and this can have the same effect as deliberate selective breeding, even when the people involved are not aware of what they are doing. Darwin points out that ancient man did this, long before the theory of selective breeding was understood and long before it was consciously practised.

The result of all these years of unconscious selection has been the creation, not only of new varieties, but of *new species* as well. Men and women have been keeping animals and cultivating plants since before the beginning of recorded history. Our domestic species descended from wild species that began to be cultivated. Now, so many small changes have accumulated that they are different species altogether. Darwin writes:

we cannot recognize, and therefore do not know, the wild parent-stock of the plants which have been longest cultivated in our flower and kitchen gardens . . . it has taken centuries or thousands of years to improve or modify most of our plants up to their present standard of usefulness to man . . .

But this is inference. What we can directly observe is the work of contemporary breeders, who go about their business with conscious purpose. Darwin notes that 'Breeders habitually speak of an animal's organization as something plastic, which they can model almost as they please.' But it is important to notice that even the most successful breeders are impotent to produce any actual changes in individual plants or animals. They must wait patiently for variations to appear, as new individuals come into being; only then can they seize upon the 'different' individuals for breeding purposes. To increase their chances of finding suitable variations, they keep large numbers of specimens, selecting the useful ones and discarding the rest.

The best breeders are able to capitalize on minute variations, which the rest of us would not even notice. Darwin, who had tried his own hand at this with pigeons, marvelled at their ability to spot 'differences which I for one have vainly attempted to appreciate. Not one man in a thousand has accuracy of eye and judgment sufficient to become an eminent breeder.' Their products make modern horticulture and animal husbandry possible.

This process, called 'artificial selection', was well known, and Darwin began by discussing it because he wanted to introduce the idea of 'natural selection' by analogy: in nature, he says, the creation of new varieties and new species

follow from the struggle for life. Owing to this struggle, variations, however slight and from whatever cause proceeding, if they be in any degree profitable to the individuals of a species, in their infinitely complex relations to the other organic beings and to their physical conditions of life, will tend to the preservation of such individuals, and will generally be inherited by the offspring. The offspring, also, will thus have a better chance of surviving, for, of the many individuals of any species which are periodically born, but a small number can survive. I have called this principle, by which each slight variation, if useful, is preserved, by the term Natural Selection, in order to mark its relation to man's power of selection.

Characteristics that Confer Advantages

In order to understand how natural selection works, and how organisms are modified by it, we must understand how an organism's particular characteristics can confer on it an advantage in 'the struggle for life'. But we immediately encounter a problem. In many instances, owing to the subtle and delicate balance that exists in nature, it may be impossible to figure out exactly why some individuals prevail over others. In nature, everything interacts. Darwin describes a heath on which fir-trees had been planted; in consequence, everything was changed: various other plants, previously unknown, began to flourish; new insectivorous birds began to live there, which of course altered the insect population drastically, and so on—'Here we see', he says, 'how potent has been the effect of the introduction of a single tree, nothing whatever else having been done.'

Or, to take a different example: Darwin points out that red clover depends on humble-bees to carry its pollen; but the number of humble-bees in a district depends on the number of field-mice (mice eat bees), and the number of mice depends on the number of cats (cats eat mice). 'Hence', he says, 'it is quite credible that the presence of a feline animal in large numbers in a district might determine, through the intervention first of mice and then of bees, the frequency of certain flowers in that district!' This interaction we can notice; but who knows how many more subtle interactions escape our attention?

Thus the sheer complexity of nature may sometimes thwart our efforts to discover why some organisms prevail while others do not. Nevertheless, we may identify with some confidence at least some of the types of characteristics that confer advantages. I will mention three of the most obvious.

1. One important type of advantage has to do with the competition for food. The food supply in an environment is never great enough to support the unlimited growth of any species; but organisms continue to reproduce at a geometric rate, even when the limits of the food supply have been reached. Therefore many of the organisms produced will die from want of nourishment, and individuals that have an advantage, no matter how small, in securing food, will be more likely to survive.

The finches of the Galapagos Islands are a good example. The islands varied in the kinds of food available to the birds. On one island, nuts were plentiful, and a wide thick beak was best for cracking them. In such an environment, a finch with a beak even a little wider or thicker than normal would have an advantage, and would tend to be more successful in obtaining food. This bird would therefore survive to pass on his characteristic beak to his descendants, while other finches, not so well endowed, would perish. On another island, where insects, not nuts, formed the most plentiful food-supply, a differently shaped beak would confer an advantage, and so the finches would develop in a different direction. From these individual differences would arise varieties, and from those varieties, species, each adapted to its local conditions. The geographical separation of the population-groups is important in allowing each variety to develop independently of the others.

The giraffe, cited by Lamarck, is another example of the same kind. Lamarck realized that, in the giraffe's specific environment, animals with longer necks would have an advantage in securing food, and so they would pass this characteristic on to their descendants, whereas those who lost out in the competition would not leave offspring. So far, so good. But he erred in thinking that the individual's longer neck was the result of his effort to reach the higher leaves. The slightly longer necks of the original individuals were chance variations, which they would have had even if it did them no good. It was simply their good luck that they happened to live in an environment in which a longer neck was advantageous.

2. Another type of advantage concerns the avoidance of predators. 'Protective coloration' is an easy example. The colour of birds varies, even within a single species; some individuals are lighter, some darker, some mottled. The result is that some are more easily visible to predators, and so are more likely to be destroyed. The ones who survive, to pass on their particular characteristics to future generations, are therefore more likely to have a colour that camouflages them against the natural background. Darwin notes that:

Grouse, if not destroyed at some period of their lives would increase in countless numbers; they are known to suffer largely from birds of prey; and hawks are guided by eyesight to their prey—so much so, that on parts of the Continent persons are warned not to keep white pigeons, as being the most liable to destruction. Hence natural selection might be effective in giving the proper colour to each kind of grouse, and in keeping that colour, when once acquired, true and constant.

It is instructive to consider what happens when an environment changes. Suppose the grouse's environment underwent some change such that the background became lighter. The darker grouse would now become more visible to the hawks, and the characteristic that previously protected the birds would work against them. The darker, not the lighter, individuals would be more likely to be devoured. But chance variation would continue to produce some lighter-coloured birds; now they would survive in greater numbers, and their individual characteristics would come to dominate the population. Before our eyes, we would see the population change colour—and this type of phenomenon has, in fact, been observed.

Protective coloration may be the most dramatic, but it is by no means the only sort of characteristics that enables organisms to avoid predators. Animals that are faster, or that can climb a tree, or withdraw into a shell, might have an advantage. So might those that emit an unpleasant odour, or have sharper hearing or keener eyesight. Plants also benefit from such devices, sometimes in unexpected ways. Darwin remarks that down on fruit is usually considered by botanists to be of trifling importance; 'yet we hear from an excellent horticulturalist, Downing, that in the United States, smooth-skinned fruits suffer far more from a bettle, a Curculio, than those with down'. Indeed, there is no limit to the types of devices that natural selection might concoct for this purpose.

3. The competition for food is typically a struggle for survival by individuals against others of their own kind: the finches all need the same

nuts or insects; there aren't enough to go around; and the birds best able to get them live. The struggle against predators is a battle for life with organisms of other kinds: if the predator wins, he eats and lives; if not, the would-be victim lives and the predator goes without food. (Thus natural selection will work continually both to improve the skills of predators *and* to improve the ability of the prey to avoid them!) But there is a third kind of struggle, not between organisms, but between organisms and the elements.

Suppose the climate of a region is growing colder. In furry animals—wolves, for example—there will be random variations in the thickness of the coat, and as the weather grows colder those with thicker fur will be more likely to survive. As a result the average thickness of the animals' fur will increase. Or imagine that the average annual rainfall in an area is dropping. The plants with slightly longer roots will fare better, and organisms needing less water will begin to replace those needing more. As the vegetation changes, of course, the population of herbivorous organisms will vary with it, and this in turn will affect the survival of the organisms that prey on *them*—again we see how interdependent are the elements of the ecological system.

Many other examples could be given, including ones that do not fit so easily within these three categories. (Variations that make organisms more resistant to disease, for example, will be preserved, while those that make them more vulnerable will be eliminated.) But these are enough to make it clear how natural selection operates, at least on the most elementary level. It is the mechanism for which the pre-Darwinian evolutionists had searched.

Causes of Modification Other than Natural Selection

Darwin held that 'Natural Selection has been the most important, *but not the exclusive*, means of modification.' To his dismay, later Darwinians ignored the qualification, and insisted that natural selection is the sole force controlling evolutionary change. But Darwin himself always regarded natural selection as only the first among several influences at work. There are, he thought, at least four others.

1. First, there is the principle of correlated variation. The parts of an organism may be interconnected so that, if one part is modified, another part must also change in consequence. Some of these interconnections are easy to understand: for example, if the body weight of an animal increases, the thickness of the leg-bones must increase as well; otherwise the legs could not support the weight. In other cases, however, the correlations may be quite bizarre and unexpected. Breeders had learned this to their regret. They had often found that they could

not breed for one characteristic without affecting others. Darwin reports that:

Breeders believe that long limbs are almost always accompanied by an elongated head. Some instances of correlations are quite whimsical: thus cats which are entirely white and have blue eyes are generally deaf . . . white sheep and pigs are injured by certain plants, whilst dark-coloured individuals escape . . . Hairless dogs have imperfect teeth; long-haired and coarse-haired animals are apt to have, as is asserted, long or many horns . . . Hence if man goes on selecting, and thus augmenting any peculiarity, he will almost certainly modify unintentionally other parts of the structure, owing to the mysterious laws of correlation.

It is the same in nature. If it were an advantage in a certain environment for a horned animal to have longer hair, natural selection would result in the hair being longer. But if Darwin's breeders were right about the correlations, these animals' horns would also become longer or more numerous—even though there is no advantage in it.

2. Darwin also emphasized that an organ adapted for one purpose may subsequently be used to serve other purposes as well. This was an important point, for it enabled him to explain how an organism can come to have a combination of characteristics that works together in unexpected ways. For example, consider the ability of birds to fly. Flight, of course, is a highly adaptive capacity; it is useful in all sorts of ways. But how could this capacity have evolved? Even if the ability to fly confers advantages, how could it be 'selected for'? The problem is that this ability depends on a whole array of characteristics—a relatively light body-weight, an aerodynamic body-shape, hollow bones, wings of a certain shape and the muscles and other anatomical parts to manipulate them, feathers, and so on. If we try to tell a story comparable to the story of the finches' beaks, or the story of the wolves' fur, we are stymied. Are we to imagine that all these characteristics appeared at once? Or that they appeared together in rudimentary form, enabling primitive birds to fly just a little, and that this small ability was refined by further variations in later generations? It is hardly believable.

But the story is believable if we can show that each of the characteristics needed for flight could have developed originally for some *other* purpose, so that later they were available, serendipitously, for flight. Feathers, for example, may have been developed originally from scales, the transition occurring because feathers were good heat-insulators. A few members of a scaly (and earth-bound) species varied slightly from their fellows, by having scales that were a tiny bit like rudimentary feathers, and this proved useful because it gave them slightly better protection from the cold. Over many generations, as this characteristic was passed on and further variations occurred, the scales became more and more like feathers, until they *were* feathers. Thus feathers originally served the same purpose as fur on other animals—it was an adaptive response to the same need. (This hypothesis is confirmed when we observe that feathers still serve this purpose for some species that never developed flight.) But feathers happened to have aerodynamic qualities that fur did not have, and so, when fleeing from predators, the feathery animals could move faster, sailing a bit from spot to spot. Thus a characteristic that was originally selected for its thermodynamic quality came to be selected for its predator-avoidance quality; and as other modifications were made, the overall result was a species capable of taking to the air.

Now you may have noticed something odd about all this. I have taken up these topics under the heading 'Causes of Modification *Other than* Natural Selection'. Yet neither of them seems contrary to the spirit of natural selection. In fact, no other agency of change has been mentioned. Correlated variations occur *because* natural selection has favoured the characteristics correlated with them. Feathers are available for flight *because* they were first naturally selected for other reasons. So why does Darwin regard these matters as requiring something more than natural selection for their explanation?

To see why, we need to distinguish two questions:

(a) whether the presence of a characteristic (e.g. longer or more numerous horns, or the ability to fly) can be explained as having been causally produced by natural selection; and

(b) whether the characteristic in question has been 'selected for' because it is *itself* adaptive.

If we concentrate on the former question, then we have little need for supplementary principles. Natural selection (together with the biochemical rules governing organic life) explains everything. But the latter question raises a different issue. If we pick out a characteristic of an organism, and ask 'Was *that* characteristic produced by natural selection because *it* benefited the organism?', the answer is not always yes. The longer and more numerous horns were not developed because *they* were adaptive, and the ability to fly was not developed because *it* was adaptive. Darwin regarded these as important departures from his main idea, and so he stressed that unqualified natural selection is not the only force that influences the development of organic life.

3. 'Sexual selection' is another principle that Darwin regarded as different from natural selection. Species are typically divided into males and females, and in order to leave offspring an individual must have a

mate. Thus, even if an individual has managed to survive to the point at which it is ready to reproduce, there is still another struggle to be faced: the competition for mates. Darwin remarks that

This form of selection depends, not on a struggle for existence in relation to other organic beings or to external conditions, but on a struggle between the individuals of one sex, generally the males, for the possession of the other sex.

Thus some characteristics will be preserved and enhanced from generation to generation, not because they enabled individuals to survive, but because they enabled them to win mates.

The 'struggle for possession of the other sex' often involves fighting, and some characteristics are obviously associated with combat: thus Darwin notes that 'A hornless stag or spurless cock would have a poor chance of leaving numerous offspring.' But in other species the competition may be more peaceful—among birds, the male with the best song or the more gorgeous plumage may be more successful in attracting the female. This has an odd consequence: it means that the female's aesthetic sense may play a part in determining what male characteristics will be passed on to future generations. Darwin says, 'I can see no good reason to doubt that female birds, by selecting, during thousands of generations, the most melodious or beautiful males, according to their standard of beauty, might produce a marked effect.' (The same, of course, may be said about the male's aesthetic sense affecting which female characteristics will be passed on. The emphasis on the determination of male characteristics is a reminder that Darwin, like others of his day, and our own, was guilty of sexist bias.)

There is one other reason Darwin thought sexual selection to be importantly different from natural selection. They can, at times, work against one another. Characteristics that are useful in attracting mates may be detrimental in the struggle for life. A stag's horns may become cumbersome, and a bird's gorgeous tail feathers may become so big that they are a hindrance in foraging for food. Thus the two forces may push development in opposite directions.

Nevertheless, as with the first two 'qualifications' to natural selection, sexual selection seems not altogether out of keeping with the spirit of the main principle. Natural selection operates to preserve characteristics that enable organisms to survive to the point at which they are able to reproduce; sexual selection operates to preserve characteristics that benefit the organism once that point is reached. Both promote the same goal: differential reproductive success. To a thinker less scrupulous than Darwin, these two principles might seem to be no more than different aspects of one overall process.

4. Finally, there is one other force that, in Darwin's view, influences the development of species, and this one really is totally separate from, and out of keeping with the spirit of, natural selection. There is, in Darwin's overall view, more than a trace of Lamarckism. In the *Origin* Darwin repeatedly says that the 'effects of use or disuse' may be inherited. In a later work, *The Variation of Plants and Animals under Domestication* (1868), he developed a theory called 'pangenesis' to explain how this happens. According to pangenesis, each of an organism's cells throws off particles called 'gemmules', that encapsulate the properties of the parent cells. The gemmules are collected in the reproductive organs, where they interact with the germ cells, and thus help to determine the character of the offspring. (Thus the nature of the 'germ cells' is not independent of the nature of the 'somatic cells'—and this, as we have seen, was an essential feature of Lamarckism.) So, if the individual's organs have been modified by 'use or disuse', the modification might affect the gemmules thrown off by it, and so the modification might be passed on to the offspring by this mechanism.

The theory of pangenesis gained few supporters. Despite Darwin's pleas on its behalf, it was largely ignored by his disciples, and today it is recognized as one of his few real blunders. No one in those days knew exactly how inheritance works, and Darwin can hardly be blamed for speculating about it. But his speculation on this point was misguided. However, the failure of pangenesis strengthened, not weakened, Darwin's overall theory. It was the only aspect of his theory that really departed from the spirit of natural selection. With its failure, natural selection emerged as an even more powerful, all-inclusive idea.

THE OPENING ROUNDS OF CONTROVERSY

When a new theory, or a new way of looking at things, is on people's minds, everything begins to be viewed in light of it. In the years immediately following publication of the *Origin*, a number of things happened that made Darwin's theory look better and better.

In 1861 the first fossil remains of the archaeopteryx were discovered. This strange animal, unlike anything previously known, seemed to be half reptile and half bird. Ten years earlier it might have attracted little notice—it would have been just another peculiar fossil. But now Darwin's theory was on people's minds, and according to his theory, birds would have evolved from reptiles. The archaeopteryx, therefore, was taken as evidence that he was right; and this evidence was all the more impressive because, at the time Darwin had conjectured this link, there had been nothing in the fossil record to confirm it.

Other discoveries followed. In the 1860s an American palaeontologist, Othniel Charles Marsh, conducted fossil-hunting expeditions into the American West. (His guide was the famous scout William 'Buffalo Bill' Cody.) On these expeditions he found the fossil remains of toothed birds with clear reptilian features; analysing them, he was able to confirm that scales had become feathers, that forelimbs had developed into wings, and to chart the other modifications that had transformed reptiles into birds. Each new discovery of this kind increased the plausibility of Darwin's view.

In the *Origin* Darwin had studiously avoided the question of human evolution. There was resistance enough to the idea that *any* species might be transformed into another, and he did not want to complicate matters needlessly by considering the emotionally charged question of man. But he did not ignore this matter entirely. At the very end of the book, he issued a short warning: he predicted that, as a result of his investigations, 'Much light will be thrown on the origin of man and his history.'

The warning was hardly necessary. Everyone who read the book drew the obvious conclusion. If birds are descended from reptiles, it is clear that humans must be descended from—well, apes. In the same year the archaeopteryx was discovered, a French traveller brought to England some stuffed gorillas he had killed in Africa. The gorillas, which had not been seen before in Europe, resembled man more than any previously known species. Ten years earlier, they would have been just another curiosity. Now the reaction was so strong that some people denounced them as frauds.

The controversy over Darwin's book was conducted with the courtesy expected of Victorian gentlemen. Even the most savage attacks were prefaced by warm acknowledgements of Darwin's other contributions to science, and personal relations between the combatants was, at least on the surface, cordial. In order to maintain the impersonal character of public disputation, it was customary that book reviews be unsigned—although, in the case of the *Origin*, everyone quickly figured out who had written what.

The Church's position was predictable. After all, it was not only the chronology of Genesis, or Archbishop Ussher's calculation, that was at issue. Christianity is, first and foremost, a historical religion. More than one Christian theologian has observed that 'Our God is a God of history.' God's relation to man has been revealed in a series of historical events, each one with deep spiritual significance—the Creation, the Fall, Atonement, and Redemption. Moreover, history has not only a beginning, in God's original act of creation, but a goal and purpose: it is leading ultimately to the establishment of God's Kingdom. The study of history, in the Christian tradition, is the study of God's interactions with man. If all this is shown to be fiction, faith loses its foundations. Darwin had attacked sacred history.

The leading clerical critic of Darwin's book was Samuel Wilberforce, Bishop of Oxford, called 'Soapy Sam' because of his oratorical gifts and his slippery style in debate. Despite the dismissive nickname, Wilberforce was a man of considerable substance. His family had been prominent in the movement to abolish slavery, which made him and Darwin allies in at least one respect. He was a decent, if undistinguished, naturalist, with good ornithological work to his credit. He was also a gentleman in the old style, unfailingly gracious, and he and Darwin got on well together personally. But Bishop Wilberforce had no doubts about the perniciousness of the evolutionist heresy. He knew that more was at stake than mere chronology. In the *Quarterly Review*, after paying tribute to Darwin's 'really charming writing', he declared:

Now, we must say at once, and openly, that such a notion is absolutely incompat-ible not only with single expressions in the word of God on that subject of natural science with which it is not immediately concerned, but, which in our judgment is of far more importance, with the whole representation of that moral and spiritual condition of man which is its proper subject-matter. Man's derived supremacy over the earth; man's power of articulate speech; man's gift of reason; man's freewill and responsibility; man's fall and man's redemption; the incarna-tion of the Eternal Son; the indwelling of the Eternal Spirit—all are equally and utterly irreconcilable with the degrading notion of the brute origin of him who was created in the image of God, and redeemed by the Eternal Son assuming to himself his nature.

Wilberforce figured prominently in the most famous public debate of Darwin's book. At a meeting of the British Association for the Advance-ment of Science, in the summer of 1860, a man named John William Draper was scheduled to read a paper discussing the new theory. The contents of his address have long since been forgotten. It was the discus-sion following the main presentation that is so vividly remembered.

The meeting resembled a convention of the most important men in Darwin's life. His old friend and mentor Henslow was in the chair. Also present were Captain FitzRoy, Joseph Hooker, and T. H. Huxley. Henslow, FitzRoy, and Hooker had of course known Darwin for many years. Huxley had only recently become intimate with Darwin, when he had volunteered to represent him in the public debates for which Darwin had little taste. Darwin was singularly fortunate in this, for Huxley turned out to be, in the words of Loren Eiseley, 'the most formidable scientific debater of all time'. Huxley had been assigned to write the review of the *Origin* for *The Times*, the single most important shaper of public opinion—for Darwin, a piece of great good luck.

Despite his formidable debating skills, Huxley was reluctant to attend the BAAS meeting. He knew that Wilberforce was going to be there, and he feared that nothing could be done to best the clergyman. Wilberforce, with his matchless style, would surely carry the day. But Huxley was prevailed upon to attend anyway.

A time-traveller eavesdropping on that meeting would probably find that what actually happened is different, at least in some details, from the stories that have been handed down. No one transcribed the discussion, and such documents as we have disagree. But this is what at least some observers say occurred.

After Draper had finished reading, and some other people had spoken, Wilberforce rose and made a speech eloquently defending religion and morality against the Darwinian onslaught. Wilberforce was, as we have already noted, a respectable naturalist, and he was present at the meeting not as a representative of the Church but in his role as Vice-President of the British Academy. He was therefore able to buttress his religious objections with scientific arguments. Nodding to Huxley, he slyly told the crowd that Darwin's champion would surely refute his logic, but that he was nevertheless obliged to speak out. In closing, however, Wilberforce made a grave tactical error. He enquired of Huxley whether he was descended from monkeys on his father's side, his mother's side, or both. Hearing this, Huxley whispered to a companion: 'The Lord hath delivered him into mine hands.' And to Wilberforce he shot back: '*I* would rather be the offspring of two apes than to be a man and afraid to face the truth.' One of the ladies present is said to have fainted—this was, after all, supposed to be a gentlemanly debate. Others in the audience applauded loudly. Lyell, who was not at this meeting, wrote a letter a few days afterwards recounting what he had heard. 'Many blamed Huxley', Lyell wrote, 'for his irreverent freedom; but still more of those I heard talk of it, and among them Falconer, assures me that the Vice-Chancellor Jeune (a liberal) declared that the Bishop got no more than he deserved.'

Captain FitzRoy then rose to speak of his association with Darwin thirty years before, and of the disappointment he now felt about the talented young man's having gone wrong. FitzRoy, now a sad figure, had brought a Bible with him; he lifted it above his head, and proclaimed that it, not men's theories, was the source of truth.

Finally, Hooker addressed the group. He spoke persuasively of Darwin's integrity and devotion to truth; and he defended the theory of natural selection as a powerful and impressive discovery. He did not ridicule those who were resisting the new theory. Instead, he sympathized with them. Whereas Huxley had been combative, Hooker was conciliatory. He concluded on just the right note:

I knew of this theory fifteen years ago. I was then entirely opposed to it; I argued against it again and again; but since then I have devoted myself unremittingly to natural history; in its pursuit I have travelled round the world. Facts in this science which before were inexplicable to me became one by one gradually explained by this theory, and conviction has been thus gradually forced upon an unwilling convert.

Lyell's friends told him that 'The Bishop had been much applauded in the Section, but before it was over the crowded section (numbers could not get in) were quite turned the other way, especially by Hooker.'

No doubt this version of events is somewhat embellished. Nevertheless, this debate, coming less than a year after the *Origin* had been published, was remembered by many of the participants as a rousing victory for the Darwinians. For Darwin, the public discussion of his theory had gotten off to a good start.

It would be a mistake to think that the controversy over Darwin's book was only a matter of the scientists versus the Church. Although Darwin was lucky to have such leading figures as Hooker and Lyell on his side, the rest of the scientists were by no means united in his support. Adam Sedgwick, the geology professor at Cambridge who had encouraged Darwin as a student, who had taken him on a three-week walking tour of North Wales, and who had praised him so highly for his work on the *Beagle*, now turned against him. Sedgwick had predicted that Darwin 'would have a great name among the naturalists of Europe'. Now Sedgwick thought it had become an infamous name. A devout man, he condemned the theory of natural selection not only as 'utterly false', but as 'grievously mischievous' because it conflicted with revealed truth. He told Darwin privately, 'I have read your book with more pain than pleasure. Parts of it I admired greatly, parts I laughed at till my sides were almost sore.' And in print he denounced the book in the strongest terms. 'Poor dear old Sedgwick,' Darwin said. 'Now I know that a man may roast another, and yet have as kind and noble a heart as Sedgwick's.' Sedgwick's reaction is a reminder of how recently leading scientists still viewed their scientific work as subordinate to the teachings of the Church.

Many of Darwin's scientific critics were much more distinguished than 'poor dear old Sedgwick'. Richard Owen, the leading comparative anatomist of his day, and sometimes called the 'British Cuvier', attacked Darwin with unusual hostility. Owen's opposition had an ironic twist.

He was himself an evolutionist—although he did not always make that clear in his attacks on Darwin—but with a different theory of how change takes place, and his biting criticism seemed motivated at least in part by jealousy of a rival. Owen's position was a hint of things to come: the majority of scientists would be convinced that Darwin was right about the fact of evolution long before they would concede that he was right about its mechanism.

The debate quickly spread beyond England. To Darwin's delight, the *Origin* was soon translated into the major European languages, and even into Japanese. In America, both his staunchest defender and his severest critic were at Harvard. Asa Gray, the botanist to whom Darwin had written the letter read by Lyell and Hooker before the Linnean Society, was the American equivalent of Huxley, championing natural selection at every opportunity. But his colleague, Jean Louis Agassiz, was Darwin's implacable enemy. Agassiz had made great contributions to geology and comparative anatomy, but he held fast to a view of nature wholly at odds with evolutionism. He believed that the order of nature is determined by the ordering of ideas in God's mind; species, therefore, are eternally fixed by God's conception of them. Today this view seems, at best, quaintly metaphysical. But pre-Darwinian biology was altogether compatible with such notions, and it was in no way a sign of scientific incompetence to hold them.

Agassiz's formal objection to Darwinism was typical of the criticisms offered by hostile scientists: Darwin, he said, has proven nothing. He has given us some clever speculation about how evolution *might* have occurred, but no solid reason to think it *did* happen that way. Darwin was especially disappointed by this kind of criticism. He expected opposition from the Church, and he expected some scientists to question his results. But he was unprepared for scientific objections to his method. His postulation of a mechanism, he thought, was no different from the standard scientific practice of assuming the minimum theoretical apparatus needed to explain observed phenomena. Darwin compared the 'unprovability' of his theory to the 'unprovability of the ether, which at that time was almost universally accepted by scientists as necessary for explaining light. In saying that he had proven nothing, the hostile scientists were insisting that *his* theory must satisfy a higher standard of evidence than the one commonly accepted for other theories.

Darwin rightly suspected that some of the scientists who resisted his view were motivated by more than strictly scientific reasons. Sedgwick, for example, was known to believe that evolutionism was a morally pernicious doctrine. In 1845 Sedgwick had written to Lyell, about Chambers's *Vestiges*, 'If the book be true . . . religion is a lie; human law is a mass of folly, and a base injustice; morality is moonshine; our labours for the black people of Africa were works of madmen; and man and woman are only better beasts!' Even Lyell, a more temperate man, had this reaction: if Chambers was right, he said, then 'all our morality is in vain'.

Darwin had delayed publishing his theory for twenty years, at least partly because he dreaded the controversy it would create. Now that the controversy was upon him, he seemed intent upon avoiding it as much as possible. Throughout the 1860s he stayed at home. He did not debate. He did not attend scientific meetings. Instead, he plunged into a study of orchids. This reluctance to enter the fray has been variously attributed to his bad health, to his distaste for quarrelling, and to his life-long reluctance to distress his wife. All this may be true. But there is another explanation that is equally plausible. Darwin was eager to defend his theory, not by winning short-term victories in public debate (which he was not very good at anyway), but by continuing to adduce evidence in its favour. He worked continually on new editions of the *Origin*—there were six before his death in 1882—adding, in each succeeding edition, more arguments and more replies to his critics' objections.

Moreover, the new books he produced in this period, while they appeared superficially to have nothing to do with his species theory, really constituted new evidence for it. In the *Origin*, one of his favourite examples of adaptation was the 'co-adaptation' of plants and the insects they rely on to spread their pollen. His book on orchids, *On the Various Contrivances by which British and Foreign Orchids are Fertilised by Insects*, published in 1862, was a study of these exquisite adaptations, support-ing his view of their nature and origins. Similar remarks can be made about his other major works of this period—*The Movements and Habits of Climbing Plants* (1865) and *The Variation of Plants and Animals under Domestication* (1868). Darwin was working to advance his theory in the best and most permanent way he could, by giving it firmer scientific support.

The anti-evolutionists were, of course, fighting in a losing cause. Before Darwin died he saw the vast majority of naturalists come to agree that evolution had occurred. But this did not mean that they auto-matically counted themselves as Darwinians. Other evolutionary schemes were proposed, and the debate came to focus on *how* change takes place. Because Darwin had not originated the idea of evolution itself (although he was certainly responsible for making it respectable), he would not be regarded by all the new evolutionists as their leader. But he knew that, once the basic idea of evolutionary change was accepted, half his battle was won. He wrote, in a journal of popular opinion,

Whether the naturalist believes in the views given by Lamarck, or Geoffrey St Hillaire, by the author of the 'Vestiges', by Mr Wallace and myself, or in any other such view, signifies extremely little in comparison with the admission that species have descended from other species and have not been created immutable; for he who admits this as a great truth has a wide field opened to him for further inquiry. I believe, however, from what I see of the progress of opinion on the Continent, and in this country, that the theory of Natural Selection will ultimately be adopted, with, no doubt, many subordinate modifications and improvements.

The question of whether natural selection is the primary mechanism of change was not finally settled in Darwin's favour until well into the twentieth century.

In the mean time, there was another pressing question to be addressed. Many of Darwin's contemporaries were ready to accept a general evolutionary scheme, but were unwilling to include man within its scope. Man had always been viewed as 'different' from the rest of nature, and not only by theologians. Virtually every thinker in the Western tradition had regarded man as set apart from other animals, by virtue of his rationality, his free will, and his moral sense. Few nineteenth-century scientists were willing to give up this exalted view of humankind, and they strove to find ways to except human beings from the laws that they now admitted must govern the rest of nature.

MAN AND THE APES

Darwin had not discussed human origins in his great book, but he knew that sooner or later he would have to face this issue forthrightly. He finally did so in two works published in the early 1870s, *The Descent of Man and Selection in Relation to Sex* (1871) and *The Expression of the Emotions in Man and Animals* (1872).

But once again Darwin had delayed—twelve years had passed between the *Origin* and the *Descent of Man*—and by the time he finally got around to applying his theory to man, others had already begun the work. Lyell had published *The Geological Evidences of the Antiquity of Man* in 1863, four years after the *Origin*, and had shown that human history stretches back into geological, not merely historical, time. In the same year, Huxley had published *Evidences as to Man's Place in Nature*, arguing that 'the structural differences which separate Man from the Gorilla and the Chimpanzee are not so great as those which separate the Gorilla from the lower apes'. In Germany, the evolutionist Ernst Haeckel had advanced the project even farther in two works, *General Morphology* (1866) and *The Natural History of Creation* (1868). Referring to the second edition of the latter book, Darwin said, 'If this work had appeared before my essay [*The Descent of Man*] had been written, I should probably never have completed it.' But it hardly mattered that these other works preceded his. Darwin was now the pre-eminent figure in this particular field, and as before, his work would be, at least for a time, the definitive statement.

Similarities between Men and Other Animals

Darwin was convinced, of course, that man is subject to the same laws that govern the rest of nature, and that we, like the other animals, are descended from more primitive forms. But the most impressive evidence of this—the fossil remains of early hominids—had not yet been discovered. Therefore, Darwin had to argue for this conclusion indirectly, by stressing the similarities between humans and the other animals. We are so much like them, he said, that if *they* have evolved, then it is only reasonable to think that *we* have evolved also, and by the same mechanism.

What were those similarities? First, like the other animals, man is subject to slight variations from individual to individual, and these variations are passed on by inheritance. Man also reproduces in greater numbers than can survive. These facts alone would be enough to clinch the case, in Darwin's view, for these are just the facts that enable natural selection to operate. But there was more. Any species with an extended range will tend to diversify; individualized, geographically separated varieties will appear. This happens with man: Africans, Eskimos, and Japanese are, to the biologist's trained eye, distinct varietal forms. Moreover, as biologists had always known, it is easy to fit man into the great classificatory scheme: he is a primate, a mammal, a vertebrate. Once these classifications are seen as related to lines of evolutionary descent, it is clear that man also belongs to a particular line of descent.

We are linked with the apes because, like them, we are primates. Every detail of our biological make-up attests to this kinship: Darwin stressed that men and apes are structurally similar, not just in gross anatomy, but in countless detailed ways—even in the details of his brain: 'every chief fissure and fold in the brain of man has its analogy in that of the orang.' Moreover, men and apes are subject to many of the same diseases; they host many of the same parasites; they react similarly to many drugs and chemical substances—including, Darwin notes, whisky.

Even more dramatic evidence of our evolutionary history is provided by the study of vestigial organs. Evolutionary change occurs slowly. Complex structures do not appear or disappear all at once; rather, they are gradually built up, modified, or eliminated. Therefore, any animal with an evolutionary history retains within it traces of that history. If man has evolved from other forms of life, there should be vestiges of those earlier forms in him today.

Such vestiges are clearly visible, and Darwin draws attention to several of them. Our ancestors had tails, and at the base of the human spine there is a structure for supporting a tail. The *os coccyx*, popularly and accurately called the 'tail bone', is the vestige of the tail itself. The vermiform appendix, which has now become a liability because it serves no good purpose but can become inflamed, rupture, and cause death, is another such vestige. It was once a pouch for storing food, useful when our ancestors dined more on plants than meat. Our wisdom-teeth have also become useless to us: so much so, in fact, that now they do not come in until around the eighteenth year, and they are frequently impacted. Dentists commonly remove them, recognizing that they are more trouble than they are worth.

Many mammals have muscles that enable them to move their ears. We still have such muscles, but we have no need of them, and consequently they have been rendered so feeble as to be useless. Darwin observes that 'The power of erecting and directing the shell of the ears to the various points of the compass, is no doubt of the highest service to many animals, as they perceive the direction of danger; but I have never heard, on sufficient evidence, of a man who possessed this power, the one which might be of use to him.'

Surprisingly, Darwin also treats our sense of smell as the enfeebled remnant of a power no longer of much use to us. By this power our ancestors could detect enemies and track prey. 'Civilized man', Darwin says, no longer has need of such skills, and this accounts for our sense of smell being poorly developed. Our skimpy body-hair is likewise the remnant of a once-rich coat.

A corollary of Darwin's theory is that, when a characteristic has become useless, it is thereafter subject to greater variation than before. The reason is that, when a characteristic no longer contributes to the individual's well-being, it is no longer subject to the controlling pressure of natural selection. We can easily observe this in non-human animals, and when we turn to man, we find the same phenomenon. Thus the length of the vermiform appendix is quite variable, as is the ability to move one's ears (a little, anyway), and similarly for the wisdom-teeth, the amount of hair on men's bodies, and so forth. Viewed in this light, it is plain that man has been produced by the same forces that shaped the rest of the natural world.

How the Transition to Man might have Occurred

The consideration of such facts makes it clear that man *has* descended from apelike ancestors; yet it still needs to be explained how such a transformation *could have* taken place. Despite the evidence, many of Darwin's readers could not comprehend how such a thing is possible. How could a non-human become a human? In order to make the idea plausible, Darwin needed to show how a series of small changes, each one plausible in itself, might add up to the overall transformation. The key events, he thought, would have been our ancestors coming to live on the ground rather than in trees, learning to walk on two legs, the refinement of the hands, and the enlargement of the brain. It might have happened like this:

As soon as some ancient member in the great series of the Primates came to be less arboreal, owing to a change in its manner of procuring subsistence, or to some change in the surrounding conditions, its habitual manner of progression would have been modified: and thus it would have been rendered more strictly quadrupedal or bipedal . . . we can, I think, partly see how he has come to assume his erect attitude, which forms one of his most conspicuous characters. Man could not have attained his present dominant position in the world without the use of his hands, which are so admirably adapted to act in obedience to his will . . . But the hands and arms could hardly have become perfect enough to have manufactured weapons, or to have hurled stones and spears with a true aim, as long as they were habitually used for locomotion, and for supporting the whole weight of the body, or, as before remarked, so long as they were especially fitted for climbing trees . . . From these causes alone it would have been an advantage to man to become a biped; but for many actions it is indispensable that the arms and whole upper part of the body should be free; and he must for this end stand firmly on his feet. To gain this great advantage, the feet have been rendered flat . . . as the hands became perfected for prehension, the feet should have become perfected for support and locomotion. . . .

As the progenitors of man became more and more erect . . . endless other changes of structure would have become necessary. The pelvis would have to be broadened, the spine peculiarly curved, and the head fixed in an altered position, all which changes have been attained by man . . . Various other structures, which appear connected with man's erect position, might have been added . . .

. . . The early male forefathers of man were, as previously stated, probably furnished with great canine teeth; but as they gradually acquired the habit of using stones, clubs, or other weapons, for fighting with their enemies or rivals, they would use their jaws and teeth less and less. In this case, the jaws, together with the teeth, would become reduced in size, as we may feel almost sure from unnumerable analogous cases . . .

. . . Therefore, as the jaws and teeth in man's progenitors gradually become reduced in size, the adult skull would have come to resemble more and more that of existing man . . .

As the various mental faculties gradually developed themselves the brain would almost certainly become larger . . .

The gradually increasing weight of the brain and skull in man must have influenced the development of the supporting spinal column . . .

As the numerous ellipses attest, Darwin's explanation is more detailed than this. But this is enough to convey the spirit of his account.

It is easy to misunderstand the point of this sort of historical speculation, and Darwin's early critics did misunderstand it. They complained that this fanciful tale proves nothing. What we want to know, they said, is not what *might* have happened, but what *did* happen, and Darwin's tale is mere conjecture. This criticism was misguided because, in the first place, Darwin was not trying to prove that human evolution had followed this particular course. Rather, he was trying to remove a certain obstacle to belief. Some people might be prevented from accepting the fact of human descent, despite the overwhelming evidence for it, because of their inability to imagine how such a thing could occur. Darwin was saying, in effect: it is not impossible; here is one way it could have happened. If it turned out that the actual historical sequence was different from what he conjectured, this would not have bothered Darwin at all.

Moreover, his conjecture was not merely fanciful. It was a plausible historical reconstruction that accounted for the known facts in an economical manner. As such, it formed part of a theory which could claim to provide the best available account of man's origins. The fact that it was not wholly 'proven' was irrelevant to its standing as a reasonable theory, to be elaborated, modified, and perfected as new discoveries are made. In science, the first step towards discovering what does happen is almost always a reasonable conjecture about what might happen.

Man's Mental Powers

Darwin was aware, however, that evidence relating only to man's physical development would not be enough to convince the sceptics. From ancient times man has thought himself special because of his higher intellectual capacities. Man is the rational animal. Any convincing account of man must explain this, the most important of his qualities— but how can reason be explained as the product of natural selection? In the opinion of many critics, this was the severest difficulty facing Darwin's theory.

Consider, for example, the tortured position of St George Jackson Mivart, an outstanding biologist, Fellow of the Royal Society, and Linnean. After publication of the *Origin*, Mivart had accepted Darwin's theory, and was an important convert because his standing as a Catholic was every bit as secure as his position as a scientist. He became a friend of Huxley's, and was, if not one of the inner circle, at least one of Darwin's firm supporters. But in the end he was unable to go all the way with the theory: in 1869 he visited a surprised Huxley to tell him that he was breaking away. Mivart then became the leader of a group of dissident evolutionists who held that, although man's body might have evolved by natural selection, his rational and spiritual soul did not. At some point God had interrupted the course of human history to implant man's soul in him, making of him something more than merely a former ape.

Darwin disagreed, of course, and to deal with this issue he adopted a two-pronged strategy. The first part of the strategy involves his materialism, and it is more implied than stated outright. Throughout his discussion, Darwin assumes that rationality (or intelligence, or language-use, or any of man's other mental powers) is an *ordinary* characteristic. It is not some sort of occult quality whose presence requires extraordinary principles of explanation. Thought, he had decided many years earlier, is nothing but 'a secretion of the brain', and so he assumes that, if the development of the brain can be accounted for by the principles of natural selection, there is no left-over problem about thought.

Secondly, Darwin simply denied the whole idea that there is something special about man's intellectual capacities. 'There is no fundamental difference', he said, 'between man and the higher mammals in their mental faculties.' Thus, he reasoned that if the intellectual capacities of other animals are produced by natural selection, and their capacities are not different in kind from man's, there is no reason to doubt that man's capacities are also the result of natural selection.

In thinking about non-humans, Darwin said, we have always underestimated the richness of their mental lives. We tend to think of ourselves as mentally complex, while assuming that 'mere animals' lack any very interesting intellectual capacities. But this is incorrect. Nonhumans experience not only pleasure and pain, but terror, suspicion, and fear. They sulk. They love their children. They can be kind, jealous, selfcomplacent, and proud. They know wonder and curiosity. In short, they are much more like us, mentally and emotionally, than we want to admit.

We should be careful not to misinterpret Darwin on this point. He did not deny that the intellectual capacities of humans are much more impressive than those of any other animal. He acknowledged that man far outdistances all other animals in linguistic ability, thought, and reason. He only insisted that the differences, impressive as they are, are matters of degree, not of kind. Those who wished to stress man's specialness were still free, on his view, to emphasize the differences of degree.

But the distinction between degree and kind was important. The issue, after all, was whether a certain *kind* of capacity could have been produced by natural selection. If this was admitted, then Darwin's battle would be won. Degrees would then be easy to explain; it would just be a matter of natural selection augmenting the characteristic, in the same way that natural selection augments any other characteristic. In Chapter 4 we shall examine this issue in greater detail.

Darwin's Disagreement with Wallace

Among those who exempted man from the general evolutionary scheme was Darwin's old friend and rival Wallace. Wallace took a view very similar to that of Mivart: he held that the theory of natural selection applies to humans, but only up to a point. Our bodies can be explained in this way, but not our brains. Our brains, he said, have powers that far outstrip anything that could have been produced by natural selection. Thus he concluded that God had intervened in the course of human history to give man the 'extra push' that would enable him to reach the pinnacle on which he now stands. Like Mivart, Wallace thought that this concession would help to reconcile religion and evolutionary theory. Natural selection, while it explained much, could not explain everything; in the end God must be brought in to complete the picture.

Darwin found this disagreement especially frustrating. Wallace was, after all, the co-discoverer of natural selection, and he was, as Darwin often remarked, a brilliant reasoner. Why should Wallace, of all people, abandon the theory just when the crucial issue was joined? In *The Descent of Man*, after a particularly telling point, Darwin wonders aloud: 'I cannot, therefore, understand how it is that Mr Wallace maintains that "natural selection could only have endowed the savage with a brain a little superior to that of an ape." '

Wallace's refusal to bring man entirely within the net of the theory is often presented as though it were just another example of religious resistance. Wallace's own subsequent history makes this a tempting interpretation. In later life he became a spiritualist, convinced that space is full of disembodied beings. (If this sounds batty, at least Wallace wasn't alone: towards the end of the nineteenth century spiritualism enjoyed a great vogue in England, and attracted a number of eminent people.) However, it is a mistake to see Wallace's position merely as an instance of religious bias. It was, in fact, the consequence of a deep scientific disagreement between him and Darwin.

Wallace, as it turns out, was a much more rigorous selectionist than Darwin. As we have already noted, Darwin allowed that some of an organism's characteristics may be produced in ways other than by natural selection. But not Wallace. Wallace believed that, wherever natural selection reigns, it reigns completely: he insisted that *every* characteristic and *every* capacity of an organism must be selected because of its own particular adaptive value.

When Wallace tried to apply this rigorous selectionism to man, he encountered a problem. He realized that the brains of so-called 'primitive' men—Australian aborigines, American Indians, black Africans, and so on—are identical to the brains of so-called 'civilized' men—the white Europeans. To his credit, Wallace was one of the few nineteenth-century naturalists who did not make the racist assumption that their brains were inferior. He saw that, even if the 'savages' did not actually compose sonatas and do calculus, they nevertheless have the mental capacity to do those things. And this created a problem. Obviously, these capacities could not be the products of natural selection, for they conferred no advantage in the struggle to survive—indeed, the savages had never even *used* such mental abilities. So he concluded that their brains, as well as our own, are not the products of natural selection—for 'natural selection could only have endowed the savage with a brain a little superior to that of an ape.'

Thus, Wallace was led to this conclusion, not simply by the desire to make a place for God, but by an overly-rigorous selectionism, together with the fact that he was not a racist.

Darwin does not seem to have appreciated the nature of Wallace's view on this point, for he never addresses Wallace's argument directly. If he had, he would have found it easy to reply. The savage's unused mental capacity is a good illustration of the idea that a characteristic developed for one purpose may later be used for a different purpose—a point that Darwin stressed many times. Man's large brain was doubtless developed originally because it was an advantage to him to make and use tools, to reason about environmental conditions, and so on. But the mental capacities that were developed for these purposes could then be used in other ways: this was the 'extra capacity' that civilized man, but not primitive man, has utilized.

Like Wallace, Darwin was no racist, and this contributed to one other important result. Darwin predicted that further discoveries in the field would reveal that early man had grown up in Africa, because that is where the apes, our closest relatives, are. This idea was resisted on racist grounds—how could Africa, of all places, be the original home of man? For decades, investigators searched in vain for evidence that man had first appeared in Europe or western Asia. But once again Darwin turned out to be right.

THE END OF DARWIN'S LIFE

When *The Origin of Species* was published, Darwin was 50 years old; when *The Descent of Man* was published, he was 62. He had eleven more years to live. He could have retired at this point, and his place in history would have been the same. But his notebooks were bulging with the observations of a lifetime; and during those final years he produced an extraordinary series of books that would add still more lustre to his reputation. *The Expression of the Emotions in Man and Animals* (1872) was a continuation of the argument that man is closely linked to the other species. Five other books, published one after the other, treated less controversial subjects, but always with an eye to their implications for the great question of descent: *Insectivorous Plants* (1875); *The Effects of Cross-and Self-Fertilization in the Vegetable Kingdom* (1876); *The Different Forms of Flowers on Plants of the Same Species* (1877); *The Power of Movement in Plants* (1880); and *The Formation of Vegetable Mould, through the Action of Worms, with Observations on their Habits* (1881).

In 1882 Darwin suffered a series of heart attacks and died. The family made arrangements to have him buried in a cemetery near Down House, and a local carpenter fashioned a plain, roughly-hewn casket, as Darwin had said he wanted. But the carpenter's simple casket was never used. A movement was quickly begun to give Darwin a grander resting-place. Such was the esteem in which he was held that three days after he died twenty members of Parliament formally proposed that he be buried in Westminster Abbey, where every British sovereign since William the Conqueror had been crowned, and where the heroes of England are traditionally laid to rest. After some hesitation, the family agreed. Wallace, Hooker, and Huxley were among the pallbearers, and a special anthem was composed for the occasion by the Abbey organist, with a text from the Book of Proverbs: 'Happy is the man who finds wisdom and getteth understanding.'

It was a fitting text. Happiness, according to Aristotle, is not a momentary sensation; it is a quality of a whole life. By any reasonable standard Darwin had an extraordinarily happy life. As a young man, he enjoyed the benefits of a warm and stable home, and a father prosperous enough to send him to whatever schools seemed best. He went on a great voyage around the world, and returned home to a scientific community ready to receive him with friendship and respect. His work was well-regarded from the beginning, but later on was acclaimed as profoundly important. When controversy erupted, he was so widely admired that it hardly touched him personally; and in any event, he had many able friends eager to take up his cause. Financially independent, he never wanted for anything that money could provide. And his marriage, to a remarkable woman, was happy to the end. His understanding exceeded that of anyone who had come before him, and as the proverb said he should be, he was happy.

As his burial in Westminster Abbey attests, Darwin's reputation at the time of his death was at a high point. As he himself wrote, the younger naturalists were accepting it as proven that 'species are the modified descendants of other species'; and many of these rising scientists were accepting his version of how the modifications take place. Only 'the older and honoured chiefs' were still opposed to evolution in any form. Darwin died confident that his view would prevail.

2　How Evolution and Ethics Might be Related

THE history of thought since Darwin has produced two noteworthy attempts to connect evolution and ethics: the first was the programme of 'evolutionary ethics' championed by Herbert Spencer; the second is the more recent proposal by some sociobiologists to reinterpret morality along lines suggested by their new science. In this chapter I will begin by reviewing these two projects, neither of which seems especially promising. Then I will describe a different and perhaps more profitable line of enquiry.

FROM SPENCER TO SOCIOBIOLOGY

A few months after publication of the *Origin*, Darwin wrote to Lyell, with obvious amusement: 'I have noted in a Manchester newspaper a rather good squib, showing that I have proved "might is right" and therefore Napoleon is right and every cheating tradesman is also right.' The Manchester editorialist thought this was a reason for rejecting Darwinism—obviously, a theory with such implications should not be accepted. Others, however, would soon take a different view. They would accept Darwin's theory, and conclude that the morality of ruthlessness was a good lesson to be learned from it.

Although it was not really fair to him, the name of Herbert Spencer was most commonly associated with this idea. Spencer, eleven years younger than Darwin, came from a free-thinking Derbyshire family and as a young man was attracted to radical causes such as the suffragist movement. He wandered from job to job, dabbled in journalism, and ended up as a kind of free-lance writer on intellectual subjects. In 1851 he published a book, *Social Statics*, advocating a version of evolutionary theory inspired by Lamarck and drawing lessons from it about the nature of happiness. As science, the work had little merit, which is not surprising considering that Spencer was largely self-educated and had not bothered to train himself in natural history. Although his theory received scant attention from serious naturalists, in the small world of British intellectuals Spencer became a 'somebody', and after the publication of *The Origin of Species*, his reputation grew.

But if Spencer was no scientist, he was a philosopher of some talent. As Darwin's influence spread, it became fashionable to think that evolutionary ideas could be adapted to explain a broad range of human phenomena, from art and religion to politics and ethics. Spencer was in the forefront of this movement. Although Darwin sometimes expressed dismay about Spencer's writings, especially the scientific portions, he generally held him in high regard. In 1870 Darwin wrote to a friend,

It has also pleased me to see how thoroughly you appreciate (and I do not think that this is in general true with the men of science) H. Spencer; I suspect that hereafter he will be looked at as by far the greatest living philosopher in England; perhaps equal to any that have lived.

Darwin was a poor prophet. Today almost no one reads Spencer. Far from being regarded as a great philosopher, university courses in the history of the subject may not even mention him at all. Nonetheless, he was an ingenious man, and he produced a series of interesting books 'applying' the theory of evolution (as well as other scientific enthusiasms of the day) to the main problems of philosophy. In their day his books made quite a stir.

Spencer's popularity was greatest in America. As early as 1864, the *Atlantic Monthly* proclaimed that 'Mr Spencer represents the scientific spirit of the age.' The president of Columbia University went even farther: 'We have in Herbert Spencer', said F. A. P. Barnard, 'not only the profoundest thinker of our time, but the most capacious and most powerful intellect of all time.' Such was the American enthusiasm for the man who sought to draw lessons from Darwinism. What earned Spencer such extravagant praise? Partly it was a general exuberance for the new 'scientific' approach to understanding human affairs. But the enthusiasm was also due to the way in which Spencer's doctrines seemed to vindicate American capitalism. 'The survival of the fittest' was quickly interpreted as an ethical precept that sanctioned cutthroat economic competition.

Capitalist giants such as John D. Rockefeller and Andrew Carnegie regularly invoked what they took to be 'Darwinian' principles to explain the ethics of the American system. Rockefeller, in a talk to his Sunday School class, proclaimed that 'The growth of large business is merely a survival of the fittest . . . The American Beauty rose can be produced in the splendor and fragrance which bring cheer to its beholder only by sacrificing the early buds which grew up around it. This is not an evil tendency in business. It is merely the working-out of a law of nature and a law of God.' Carnegie, who became a close friend of Spencer's, was equally rhapsodic: in defending the concentration of wealth in the hands of a few big businessmen, he proclaimed that 'While the law may sometimes be hard for the individual, it is best for the race, because it ensures the survival of the fittest in every department.' Rockefeller's and Carnegie's understanding of natural selection was only a little better than that of the Manchester editorialist.

Some of Spencer's less cautious writings encouraged this interpretation. He was a robust champion of self-reliant individualism and an advocate of free-enterprise economics. Moreover, he was delighted to have the friendship of such men as Carnegie, who entertained him lavishly when he visited America. But he was not a crude thinker. In his works on ethical theory, Spencer took a sober and cautious approach. In those works, we do not find facile 'deductions' of economic individualism from evolutionary principles. Nor do we find the vulgar slogans that were associated with his name: he did not say that 'Whatever gives one an advantage in the struggle for life is right'; nor did he try to make 'the survival of the fittest' into some sort of ethical maxim. His view was more sophisticated than that, and his theoretical treatises were careful works that could be read, and sometimes admired, by serious philosophers.

Spencer's *Data of Ethics* appeared in 1879. In it he begins by announcing the 'pressing need' for 'the establishment of rules of conduct on a scientific basis', because the advent of Darwinism had destroyed the old verities. 'Now that moral injunctions are losing the authority given by their supposed sacred origin,' he says, 'the secularization of morals is becoming imperative.' Of course, Spencer thought he knew how to do this.

Ethics, Spencer said, could be defined as the area that 'has for its subject-matter that form which universal conduct assumes during the last stages of its evolution'. Our conduct, like everything else about us, has evolved; and Spencer assumed that as a result of this evolution humans have now achieved 'higher' forms of behaviour. Of course, this untroubled use of such notions as 'higher' and 'lower' was very un-Darwinian. Darwin would never have spoken of 'the last stage' of evolution, as though it were a process that terminates in some sort of final perfection. Spencer seems not to have understood, or at least he did not accept, Darwin's point that adaptations are not 'directed' to any particular end. There is no 'more evolved' or 'less evolved' in Darwinian theory; there are only the different paths taken by different species, largely, but not entirely, in response to different environmental pressures. Natural selection is a process that, in principle, goes on forever, moving in no particular direction; it moves this way and that, eliminating some species and altering others, as environmental conditions change.

But Spencer began as a Lamarckian, and never shook off the Lamarckian notion that evolution is driven by an internal impulse towards 'higher forms'. So for him the crucial question was: Towards what end is evolution aimed? In what direction does it inevitably take us? His answer was that all creatures, including humans, have evolved patterns of behaviour that serve to increase the length and comfort of their lives. (He mentions, but does not emphasize, that evolution also favours conduct that increases the number of one's progeny.) The 'highest' form of conduct is, therefore, the conduct that is most effective in achieving these goals. What type of conduct is this? According to Spencer, the most effective conduct is the co-operative behaviour of people living together in 'permanently peaceful societies'.

Spencer is not clear about why social living is so important from an evolutionary point of view, but by emphasizing it he was able to account for one of the central features of morality, namely obligations to other people. Ethical conduct is, at least in part, unselfish conduct, and so any plausible account of ethics must explain the basis of other-regarding obligations. By conceiving of 'fully evolved conduct' as the conduct of people in *communities*, Spencer made a place for the duty of beneficence. At any rate, whatever his reason for introducing this notion, he says that 'the permanently peaceful community' represents 'the limit of evolution' as far as conduct is concerned, and so he concludes that it is 'the subject-matter of ethics'.

Having argued this to his satisfaction, Spencer then turns to the analysis of 'good and bad conduct'. He contends that 'good conduct' can be understood in the same way that we understand the notion of a 'good knife' or a 'good pair of boots'. (Socrates was also fond of this type of comparison.) Knives have a purpose; they are used for cutting. A good knife, therefore, is one that cuts easily and efficiently. Similarly, to discover the meaning of good conduct, we may ask what conduct is for, and then what type of conduct serves this purpose best. But this, it turns out, is a matter he has already settled: the purpose of conduct is to increase the length and quality of one's life and to secure offspring. This is the type of conduct that evolution produces. Therefore, Spencer concluded,

The conduct to which we apply the name good, is the relatively more evolved conduct; and bad is the name we apply to conduct which is relatively less evolved . . . Moreover, just as we saw that evolution becomes the highest possible when the conduct simultaneously achieves the greatest totality of life in self, in offspring, and in fellow-men; so here we see that the conduct called good rises to the conduct conceived as best, when it fulfills all three classes at the same time.

In summary, then, Spencer's argument seems to have been this:

[handwritten: Same not all]

1. The behavioural characteristics of a species are among the characteristics that are shaped by the evolutionary process.

[handwritten left margin: black widow – praying man? fruit fly start life?]

2. Evolution favours conduct that tends to lengthen individual lives, increase the quality of life, and, not coincidentally, secure offspring. 'More evolved' conduct is conduct that does this job better.

3. Moreover, the *purpose* of conduct is to achieve the goals of longer and better life and more offspring. *[handwritten: not at all.]*

4. These goals are best achieved when people live together, co-operating with one another, in peaceful communities.

5. It follows that the 'highest' or 'most evolved' form of conduct is conduct that creates and enhances 'permanently peaceful communities'.

6. Therefore, good conduct may be defined as conduct that achieves these goals, while bad conduct is conduct that frustrates these goals. Or, as Spencer says, 'good conduct' is 'more evolved conduct'.

Here, then, we have an ambitious, if still somewhat sketchy, theory of ethics, explicitly based on evolutionary ideas. What was to be made of it? For a while, it seemed to be a viable theory, at least no worse than other contenders in the field, and having the considerable advantage of connecting ethics with the latest scientific theory about human nature. But Spencer's popularity was short-lived. For philosophers, the publication of G. E. Moore's *Principia Ethica* in 1903 sounded its death knell.

Moore and the Naturalistic Fallacy

Moore was a Cambridge philosopher at the beginning of what was to be a long and distinguished career. His first book would become a classic, not so much for its positive claims as for its style of argument, its redefinitions of philosophical questions, and its sharp criticisms of familiar views. One of Moore's chief contentions was that all naturalistic theories of morality commit a certain mistake, which he called 'the naturalistic fallacy'. Moore used Spencer's view to illustrate how theories fall into this error. Only a dozen pages were devoted to Spencer, but Moore's charges seemed, to many readers, unanswerable. His arguments were all the more persuasive because Moore's reading of Spencer was balanced and fair, and because Moore was himself an admirer of Darwin—he was no anti-scientific yahoo, out to buttress traditional morality. After Moore's demolition, Spencer's view would seem hopelessly naïve. But it was not merely Spencer's specific formulations that were found to be defective. If Moore was right, the natural sciences, including evolutionary biology, were simply irrelevant to ethics. Readers of *Principia Ethica* would come away with the conviction that Spencer had been right about at least one thing: the foundations of ethics needed to be rethought. But most would also be convinced that 'evolutionary ethics' was a fundamentally confused idea that should have no place in the rethinking.

Taking some liberties, Moore's central argument can be stated briefly. The 'naturalistic fallacy' is committed by any theory that seeks to define ethics in naturalistic terms. Ethics has to do with what is good or right—in other words, with what *ought to be* the case. Naturalistic theories identify goodness or rightness with 'natural' properties of things—in other words, with facts about what *is* the case. But that is always a mistake. The naturalistic fallacy, therefore, is the fallacy of confusing what ought to be the case with what is the case. Spencer's theory is an example. Spencer holds that 'good conduct' is the same as 'relatively more evolved conduct'. But when we think about it, we can see that 'good' and 'relatively more evolved' are quite different notions. Whether something is good is a matter of evaluation; whereas whether something is relatively more evolved is a matter of fact. The two are not the same, and so Spencer's theory fails.

Moore's discussion of the 'naturalistic fallacy' was reminiscent of David Hume's dictum that we cannot derive 'ought' from 'is'—in fact, many commentators have opined that Moore's point was nothing more than a restatement of Hume's famous observation. In 1739, almost exactly 100 years before Darwin discovered natural selection, Hume had written in his *Treatise of Human Nature*:

In every system of morality, which I have hitherto met with, I have always remarked that the author proceeds for some time in the ordinary way of reasoning, and establishes the being of a God, or makes observations concerning human affairs; when of a sudden I am surprised to find, that instead of the usual copulations of propositions, *is*, and *is not*, I meet with no proposition that is not connected with an *ought*, or an *ought not*. This change is imperceptible; but is, however, of the last consequence. For as this *ought*, or *ought not*, expresses some new relation or affirmation, 'tis necessary that it should be observed and explained; and at the same time that a reason should be given, for what seems altogether inconceivable, how this new relation can be a deduction from others, which are entirely different from it.

Paraphrasing Hume, with an eye to Spencer's theory, we might say: Spencer proceeds in the ordinary way of reasoning, making observations concerning human affairs; and establishes that our conduct has evolved in a certain way, and that it *is* the case that we behave in these ways; but then, imperceptibly, he begins to say that this is the standard of how we *ought* to behave. But this is a new affirmation, which cannot be a deduction from the first, being entirely different from it. Or, put more plainly, the point is that the proposition 'X is good conduct' simply does not follow from the proposition 'X is more evolved conduct', and it is a mistake of logic to think that it does.

Spencer, a bright man, had known that this sort of objection might be raised, and he had tried to answer it in advance. By his reckoning, the crucial problem was this: life-prolonging and life-enhancing conduct will be regarded as good if one thinks that life is worth living. An 'optimist', who thinks life worth living, will accept Spencer's argument (or so Spencer says), while a 'pessimist', who doubts this, will have little reason to accept the argument.

Spencer attempts to settle the matter by suggesting that pessimism is based on a faulty estimate of the amount of suffering that life contains. Why should anyone think life is not worth living? This will be sensible only if life holds more pain than pleasure. The pessimist believes that it does; the optimist, that it does not. But, he says, this means that the pessimist and the optimist *agree* that pleasure and pain are the ultimate standards of reference. Spencer then announces that he himself holds pleasure and pain to be the ultimate standard:

Thus there is no escape from the admission that in calling good the conduct which subserves life, and bad the conduct which hinders or destroys it, and in so implying that life is a blessing and not a curse, we are inevitably asserting that conduct is good or bad according as its total effects are pleasurable or painful.

But this attempt to avoid Hume's problem falls short of the mark. For one thing, Spencer has now radically shifted ground. He is now asserting a very different sort of moral theory, Hedonistic Utilitarianism. Good and bad are no longer identified with what is more or less evolved; rather, they are identified with what is productive of pleasure or pain. In the new theory, the supposed 'facts' about the evolution of conduct can serve only the subordinate role of telling us what sort of behaviour does, or does not, produce pleasure. Since there are other, more plausible ways to determine this, the references to evolution are hardly needed at all. Moreover, Spencer is apparently asserting the new theory in a form that leaves it vulnerable to the very objection he was struggling to overcome. If the identification of 'good' with 'more evolved' commits the naturalistic fallacy, then so does the identification of 'good' with 'productive of pleasure'.

Moore's discussion of the naturalistic fallacy was, however, more than just a rehash of Hume's point about 'is' and 'ought'. Moore produced a new and independent argument, which has been dubbed the 'open-question argument', to demonstrate that naturalistic definitions of goodness must always be mistaken. The open-question argument goes like this. First, we note that any naturalistic definition of 'good' can be expressed in the following form:

D: 'X is good' means 'X has the property P.'

Then, we formulate two questions:

A: X has P, but is it good?
B: X has P, but does it have P?

Now the open-question argument is simply this:

If D is correct, then A and B have the same meaning.
But A and B do not have the same meaning.
Therefore, D is not correct.

And the reason A and B do not have the same meaning is that A is an 'open question' while B is not.

As Moore showed, this sort of argument can be deployed against Spencer's proposed identification of 'good conduct' and 'more evolved conduct' with good effect. Consider the questions:

A: This conduct is more evolved, but is it good?
B: This conduct is more evolved, but is it more evolved?

The first question is an 'open question'; the second is not. But if Spencer's theory were correct, they would be the same question. Hence, Spencer's theory is not correct. More generally, Moore concludes, 'good' cannot be identified with *any* of the properties investigated by the natural sciences, neither those of evolutionary biology, nor any other. Any such identification would run foul of this argument.

But does this argument really refute Spencer? Looking back, we can now see that it has less force than Moore thought. Moore construed Spencer's view as offering a *definition* of 'good conduct'—that is, he construed Spencer's thesis as a thesis about the meaning of these words. This was reasonable, since, as we have seen, Spencer had phrased his thesis as a thesis about words: he said, 'The conduct to which we apply the *name* good, is the relatively more evolved conduct; and bad is the *name* we apply to conduct which is relatively less evolved.' It is possible, however, to construe Spencer's view differently, as a claim about *what is in fact* good conduct. On this alternative reading, Spencer was offering a criterion, not a definition, of good conduct. If so, the open-question argument would not work against it.

To make the point clearer, compare Spencer's thesis to the following example. Suppose someone says: a good automobile is one that is safe, reliable, comfortable, and gets good fuel mileage. If this is intended as a definition of the meaning of the words 'good automobile', it is mistaken; we cannot define an evaluative word such as 'good' in purely factual terms, at least not in this way. Nevertheless, what is being said here is sensible, and probably true, if we take it as a claim about what properties of automobiles make them good. Construed in the latter way, the statement does not involve an is–ought confusion, nor is it vulnerable to the open-question argument. Similarly, if Spencer's thesis is interpreted as setting forth criteria of 'good conduct', it also escapes these charges. Such a criterion might be criticized on other grounds, but at least it would not be vulnerable to Moore's arguments. Spencer left himself open to Moore's criticism because he did not distinguish definitions from criteria—it is a distinction that apparently he did not notice. Moore was none too clear about the matter himself. In those days, the philosophy of language was not very far advanced.

Nevertheless, Moore was right to reject Spencer's view. Spencer's theory was not one that Darwinians could accept. Moore pointed out (what we have already observed) that Spencer's understanding of evolution was inconsistent with Darwinian notions. Darwin tried to avoid such terms as 'higher' and 'lower' when referring to stages of development—a distinctive feature of his theory was its denial that evolutionary change is associated with any purpose or 'direction'. There is no advancement and no regression; there is only change. Spencer's ethical theory, on the other hand, depended on a Lamarckian stance, on seeing some conduct as 'more evolved' than other conduct. In this way he sneaked in an evaluative element that is alien to Darwinian conceptions. As Darwin clearly recognized, we are not entitled—not on evolutionary grounds, at any rate—to regard our own adaptive behaviour as 'better' or 'higher' than that of the cockroach, who, after all, is adapted equally well to life in its own environmental niche. Natural selection favours creatures whose conduct enables them to win the competition to reproduce. Not only human behaviour, but the behaviour of countless other species, has this result. If Spencer had accepted this fundamental point, his theory would never have been conceived.

Moore's book was tremendously influential. Coming just after the turn of the century, it defined the problems that moral philosophers were to discuss for the next six decades. Evolutionary ethics was now removed from the philosophical agenda; and soon the independence of ethics from *all* the sciences would become an article of faith. By 1903, the year *Principia Ethica* was published, Spencer's books had sold a phenomenal 368,755 copies in America alone. But the vogue was over. It seems somehow fitting that, in that same year, Spencer died.

Bergson

The failure of Spencer's project did not mean, of course, that all philosophers would immediately lose interest in Darwin. Many continued to pay verbal homage to him. Mostly, however, it was just lip-service. They would say that Darwin's theory was important, but then make little use of it. In 1910 John Dewey wrote a laudatory essay with the title 'The Influence of Darwin on Philosophy'. Dewey, a philosophical naturalist, was delighted that Darwin had provided a naturalistic understanding of human origins. But, Dewey assured his readers, Darwin's work had a broader significance: 'The *Origin of Species*', he wrote, 'introduced a mode of thinking that in the end was bound to transform the logic of knowledge, and hence the treatment of morals, politics, and religion.' Reading such words, we might expect Dewey to tell us *how* Darwinism would transform morals, politics, and religion. But, after having heaped such lavish praise on the new 'mode of thinking', Dewey had little to say about its precise implications.

One philosopher who did pay more than lip-service to evolutionary ideas was Henri Bergson, the French thinker and Nobel-prize winner who had been born in the year *The Origin of Species* was published. Like Spencer, Bergson believed that an evolutionary outlook was essential to understanding virtually all human phenomena; and, again like Spencer, he enjoyed a great vogue among the intellectuals.

Bergson's *Creative Evolution* was published in 1907, and was immediately hailed as a Great Work. Upon receiving his copy of the book, William James wrote Bergson a letter filled with the most extravagant praise, beginning: 'O my Bergson, you are a magician, and your book is a marvel, a real wonder in the history of philosophy.' Like Darwin's praise of Spencer, James's tribute to Bergson today seems merely quaint. It is not just that no one reads Bergson any more—that doesn't matter; great works have often enough been neglected by subsequent generations. More importantly, Bergson can now be seen to have been fighting a rearguard action *against* the new biology and its naturalistic outlook. Appearances to the contrary, he was not the champion of Darwinism; he was its last great philosophical opponent.

Bergson was an evolutionist, but not a Darwinian: he believed that Darwin's theory was inadequate to explain the evolutionary process. Darwin had said that natural selection operates on chance variations that affect random parts of organisms. Bergson could not see how this was possible, because complex organisms are composed of interdependent parts that work together in subtle ways. If there is a variation in only one part of an organism, he reasoned, the delicate balance will be upset and the organism will be destroyed. Therefore, Bergson concluded that

Darwin's view cannot be right: the evolutionary process must be the result of various changes taking place *in concert* throughout organisms. Moreover, Bergson thought, natural selection alone cannot account for the fact that evolution produces organisms of increasing complexity.

Bergson was ready to set things right with his own version of how evolutionary change takes place. He proposed that we should recognize a new principle at work, which he called the *élan vital*, or 'life force'. The *élan vital* is a 'current of consciousness' that permeates living bodies and determines, in some obscure manner, the course of their evolution. It accounts for the co-ordinated changes in several character-istics that must occur if species are to be transformed without being destroyed, as well as explaining the drive towards increased complexity. And at the same time, reference to the *élan vital* would also help the philosopher to explain other, more spiritual matters, such as the nature of consciousness. Bergson even hints that another name for the *élan vital* might be God.

It seems clear that Darwin would have little trouble answering Bergson's specific objections. Darwin could agree, for example, that random variations affecting parts of complex organisms are often destructive, because they ruin the interactions that enable the whole organism to function. In fact, *helpful* variations might be exceedingly rare. Darwin thought this was obvious, and far from being an objection to his theory, he saw it as fitting well with his view that many more creatures perish than survive. Moreover, helpful variations were thought by Darwin to be, almost always, *slight* modifications that confer *small* advantages. That is one reason why natural selection is a slow process that might take centuries to show perceptible results. Bergson apparently did not understand this, and, judging from his fulsome praise, neither did William James. That such men could have missed the point so completely, almost a half-century after the *Origin*, only under-scores how long it takes for a revolution in thought to be assimilated. Moreover, one must remember that at this time even the scientists did not agree on the merits of Darwin's theory: while the fact of evolution was commonly accepted, the central place of natural selection in explaining how change occurs was not firmly established until the emer-gence of the 'new synthesis' in the 1930s. During the early decades of the century Lamarckian notions were still respectable.

But Bergson was a philosopher, not a biologist, and the enemy for him was a philosophical idea: mechanism, the thought that life and conscious-ness can be explained in the same terms as inanimate nature. Bergson saw the theory of natural selection as mechanistic, as leaving out of account any element that might explain what is special about the nature and purpose of living beings. Lamarck's theory, with its 'internal strivings', was, for this reason, viewed as superior. Mechanism seemed, then and now, a cold and atheistical philosophy. From this perspective, Irwin Edman's judgement seems correct: much of Bergson's popularity seems to have been a matter of 'the religious liberals welcoming a philosopher who seemed to have found critical circumvention of mechanistic science and a new and poetic support for belief in God.'

Sociobiology

During the 1930s and 40s Darwinism—or, more precisely, the 'new evolutionary synthesis', with natural selection as its centre-piece—was being established as the reigning orthodoxy in biological science. The primary concern of the evolutionists was, of course, to account for the anatomical, physiological, and morphological characteristics of organisms, as well as their geographical distribution. But there was also a growing interest in trying to explain *behavioural* characteristics by the same principles. As this project was pursued, biology was once again brought close to the territory of ethics.

The application of evolutionary principles to the explanation of behaviour requires the assumption that an organism's disposition to behave in certain ways, no less than the colour of its skin or the size of its wings, is a product of its genes. This assumption is reasonable enough, especially if the behavioural pattern is consistent throughout a species and if it appears to be instinctive rather than learned. An evolutionary account begins by asking why natural selection favours genes associated with particular forms of behaviour. What advantage in the struggle for survival is conferred by the tendency to behave in a given way?

It is easy to see how such explanations might work. An animal that grooms itself, removing parasites, obviously stands a better chance of surviving than one that allows the parasites to gnaw away. Thus the genes that produce grooming behaviour will be passed on to future generations, preserved by the same mechanism that ensures the preservation of any gene that is associated with a beneficial characteristic.

But how far can such explanations be extended? Grooming behaviour is a simple but unexciting example. Could the same strategy of reasoning be used to account for more interesting types of conduct, such as aggressiveness, male dominance, and 'moral' behaviour such as altruism? And—the big question—can we explain the *human* forms of such behaviour in such terms? If we could explain human behaviour in this way, a whole new type of understanding would suddenly become available. The underlying rationale of hitherto mysterious tendencies would be revealed; and we would see, perhaps, that aspects of our behaviour which we previously thought were matters of free choice are really the products of deep, genetically controlled forces.

This idea spread during the 1950s, and was popularized during the next decade in a series of best-selling books. Desmond Morris's *Naked Ape*, Lionel Tiger's *Men in Groups*, and Robert Ardrey's *Territorial Imperative* all advanced the same seductive thesis: human social behaviour, being the product of evolution, may be explained in the same way as the comparable behaviour of other animals who have travelled the same evolutionary path. These books were full of eye-catching claims. Ardrey, for example, argued that when humans go to war they are only acting out the same drama as other animals who stake out and defend territory. Humans may fabricate all sorts of other reasons for what they do, but these are mere rationalizations. In reality their conduct is fixed by the territorial imperative built into their genes—genes that are, after all, only slightly modified versions of the genes of other mammal species.

Such sensational claims were quickly denounced by more cautious investigators. Ardrey and the others were branded in the professional journals as bad popularizers whose speculations were unsupported by hard data. If the case for the new approach was to be made out, more careful research was needed. Some investigators, such as Konrad Lorenz in *On Aggression* (1966) had done a better job. But it was the publication of Edward O. Wilson's *Sociobiology: The New Synthesis* in 1975 that promised to raise the discussion to a new level.

Wilson, a Harvard professor, was the author of highly respected work applying evolutionary principles to the understanding of insect societies. Now he proposed to incorporate this work into a new science, sociobiology, which would be 'the systematic study of the biological basis of all social behaviour'. Much of Wilson's book dealt with non-human behaviour and was relatively uncontroversial; indeed, it drew high praise from most reviewers. The final chapter, however, was another story. Entitled 'Man: From Sociobiology to Sociology', it echoed Morris, Tiger, and Ardrey by arguing that some of the most troubling aspects of human social life are inescapable features of our human (biological) nature.

Wilson's discussion of male dominance caused immediate controversy. He pointed out that in 'primitive societies' women are controlled and exchanged by men, who are invariably dominant. The males are the hunters, the warriors, and the decision-makers. Females are submissive and do what they are required to do by the males. He then went on to compare this to social practices in modern industrial societies:

The populace of an American industrial city, no less than a band of hunter-gatherers in the Australian desert, is organized around [the nuclear family]. In both cases the family moves between regional communities, maintaining complex ties with primary kin by means of visits (or telephone calls and letters) and the exchange of gifts. During the day the women and children remain in the residential area while the men forage for game or its symbolic equivalent in the form of money. The males cooperate in bands to hunt or deal with neighboring groups.

Where there is such uniformity, Wilson thinks, there must be genetic control. Writing in the *New York Times Magazine*, he made this explicit: 'In hunter-gatherer societies, men hunt and women stay at home. This strong bias persists in most agricultural and industrial societies and, on that ground alone, appears to have a genetic origin.'

Something quite alarming had happened: what had been announced as a scientific study of human behaviour had seemingly turned into a defence of the social *status quo*. The message was clear: human social institutions such as male dominance—the men in charge, the obedient women at home caring for the babies—are inescapable features of human life, which we can no more avoid than chimpanzees can avoid the forms of their social life. Reformers who think otherwise are going against human nature itself. Other sociobiologists took up this theme, which become a staple of the literature: David Barash assured his readers that there is a 'biological basis of the double standard', while Pierre van den Berghe went so far as to urge that there must be no more 'kow-towing to feminism' on this issue.

As might be expected, such pronouncements were dismissed by critics as reactionary politics masquerading as science, while the sociobiologists replied that the critics were refusing to face hard realities. Humans are animals, too, they said; and if the behaviour of other animals is to be explained as the product of their evolutionary histories, by what right do we exempt human behaviour from the same scrutiny? No one is upset by the observation that male apes are naturally dominant. To refuse to consider a similar hypothesis about humans only betrays that we have not yet learned Darwin's lesson.

The central question, as the better critics realized, is one of science, not politics. The criticism worth taking seriously is not that the sociobiological account of male dominance is offensive; rather, it is that the account is woefully unsubstantiated. While there has been a lot of careful work done on insects and other animals—much of it done by Wilson himself—there has been no comparably detailed research on the more complicated case of human beings, on which comparable conclusions could be based. Sociobiological speculations about human nature have been based on impressions, facile generalizations, and hazy analogies. In this respect, Wilson and his followers do not seem to be much of an improvement over Ardrey, Tiger, and Morris. Therefore, the complaint goes, the conclusion about their pronouncements must be, at a minimum: not proved.

Stephen Jay Gould has accused Wilson and his followers of a more specific mistake. They fail to distinguish, he says, the plausible notion of biological *potentiality* from the much more troubling notion of biological *determinism*. What is obviously true is that our genes establish the range within which our behaviour and social institutions must fall. We would have very different lives if we photosynthesized, or had the life cycle of an insect. The fact that we do not have such lives is undoubtedly the result of our having particular kinds of genes. 'The range of our potential behaviour is circumscribed by our biology', says Gould. Everyone can agree that, in *this* sense, our behaviour is under the control of our genes.

However, it is altogether a different matter to say that our genes determine which specific behaviours, from within the available range, we will adopt. Is there a specific gene for male dominance? (Or, if male dominance is not a unitary thing, are there specific genes for the behaviours that comprise male dominance?) There are two possibilities: (1) male dominance might be only one among several forms of social organization consistent with our genes; that is, it is 'within the range' our genes permit; or (2) male dominance might be specifically mandated by our genes. Once these alternatives are firmly distinguished, Gould says, we can see that the available evidence supports only the first. 'What is the direct evidence for genetic control of specific human social behaviour? At the moment, the answer is none whatever.'

Nevertheless, although the primary issue here is scientific, it is clear that ethical and political issues are never far off-stage. Wilson was keenly aware of the ethical implications of his 'new science', and rather than trying to distance the scientific project from them, he offered a radical proposal for combining the two. Sociobiology, he proclaimed, will take over the territory previously occupied by moral philosophy. In the opening sentences of *Sociobiology*, he chides 'ethical philosophers who wish to intuit the standards of good and evil', but who do not realize that their moral feelings really spring from the hypothalamus and the limbic system. Biology, not philosophy, he says, explains ethics 'at all depths'. Later in the same work he suggests that 'The time has come for ethics to be removed temporarily from the hands of the philosophers and biologicized.'

This 'biologicizing' of ethics apparently has three parts. First, it involves the recognition that moral judgements are products of biological causes—the operation of the hypothalamus and the limbic system—and the corollary realization that they should not be regarded as occult truths known through moral intuition.

Secondly, sociobiology is to provide an answer to the problem of altruism, which has worried evolutionary thinkers since Darwin. Moral behaviour is, at the most general level, altruistic behaviour, in which an individual acts for the good of others even at some cost to himself. The problem is that we would expect altruistic behaviour to work against individual survival—the altruist increases the chances of others' surviving, by helping them, while at the same time decreasing the chances of his own survival, by giving something up. Therefore we would expect natural selection to eliminate any tendency towards altruism. Yet evidently it does not, for we find that not only humans but other animals behave altruistically all the time.

The key to solving this problem, which Wilson calls 'the central theoretical problem of sociobiology', was announced by W. D. Hamilton in some famous papers published in 1964. Hamilton's idea was based on the observation that many individuals are genetically similar to one another—typically, one shares half the genes of one's siblings, one-eighth the genes of one's cousins, and so on. Therefore, acting in such a way as to increase the chances of a genetically similar individual's surviving is a way of increasing the chances of one's own genes being passed on to later generations. This being so, we would expect natural selection to favour a tendency to altruism towards one's near kin. This fits well with the phenomenon of altruism as we commonly observe it: individuals do behave far more solicitously towards their relatives than towards strangers.

The theory of kin selection is a significant contribution to our understanding of morality, and I will have more to say about it later, in Chapter 4. For now, it is enough to observe that we have at least the beginnings of a biological theory of other-regarding behaviour, which, on Wilson's view, may usefully replace earlier philosophical and religious theories of the phenomenon. Man is a moral (altruistic) being, not because he intuits the rightness of loving his neighbour, or because he responds to some noble ideal, but because his behaviour is comprised of tendencies which natural selection has favoured.

The third and final stage of Wilson's 'biologicizing' of ethics comes when we adjust our ethical judgements to fit the realities revealed by sociobiological analysis. If it is in our genes to help our kin, but not to care about distant strangers, then it is pointless to espouse an ideal of universal altruism. Similarly, if male dominance is in our genes, the feminist cause is hopeless and teaching girls they should be independent and aggressive will only make them miserable. However, it should be emphasized that not all such 'results' are so conservative: Wilson also suggests that homosexuality may be the product of selective pressures—having a certain number of childless individuals who serve as helpers might be an effective competitive strategy—and if so, then it is wrong to condemn homosexuality as 'unnatural'.

Setting aside strictly scientific doubts about the validity of human sociobiology, what is to be made of all this? Some moral philosophers have greeted Wilson's proposals warmly; others, however, have been sceptical. I am among the sceptics, although, as will become clear, it is not because I doubt the moral relevance of evolutionary ideas.

One problem is that, while sociobiological results may be important for moral deliberation, they are important in a way that is different from what Wilson and his followers suggest. Suppose it *were* true that male dominance is an unavoidable consequence of human nature. It would not follow that the feminist analysis of its evils is false. Feminists might still be right that women's lives are impoverished when they are consigned to an inferior social status. What would follow, perhaps, is that male dominance is ineradicable. But that would only be like discovering that a dread disease is forever incurable. We might have to live with that knowledge, but we surely would not be forced to think it a good thing. Nor would we have to cease our efforts to ameliorate the suffering of the disease's victims. Similarly, we could continue to regret male dominance, and we could go on trying to minimize its effect—by continuing to extend the legal protection of women's rights, by insisting that they be paid equal wages for equal work, and so on. Nothing in sociobiology could imply otherwise.

This point is related to a deeper and more general difficulty with the idea of sociobiology's 'replacing' moral philosophy. Imagine that someone proposed eliminating the study of mathematics, and replacing it with the systematic study of the biological basis of mathematical thinking. They might argue that, after all, our mathematical beliefs are the products of our brains working in certain ways, and an evolutionary account might explain why we developed the mathematical capacities we have. Thus 'mathobiology' could replace mathematics. Why would this proposal sound so strange? It is not because our mathematical capacities have no biological basis; nor is it because it would not be interesting to know more about that basis. Rather, the proposal is strange because mathematics is an autonomous subject with its own internal standards of proof and discovery. Consider the Fundamental Theorem of Algebra, which we know to be true because of Gauss's proof. 'Mathobiology', if it existed, could add nothing to our understanding of the theorem or the proof. It would be irrelevant to determining whether the proof is valid or invalid, because that is something that can be established only within the framework of mathematics itself.

The deep reason for resisting the substitution of sociobiology for ethics is the conviction that ethics, like mathematics, is (as Thomas Nagel puts it) 'a theoretical inquiry that can be approached by rational methods, and that has internal standards of justification and criticism'. This means, first, that the observation that our moral capacities are connected with the operation of the hypothalamus and the limbic system is irrelevant to ethics in the same way that the observation that our mathematical capacities are connected with other parts of the brain is irrelevant to mathematics. Moreover, it means that particular ethical issues—such as whether male-dominated social arrangements are desirable or undesirable—are to be determined by the application of rational methods, and standards of criticism and justification, that are internal to ethics itself. That is why sociobiology can no more tell us whether sexist practices are a good thing than mathobiology could tell us whether Gauss's proof is valid. Although it might make significant contributions to our understanding of moral phenomena, the idea that sociobiology can explain ethics 'at all depths' is, for this reason, mistaken.

ARE HUMANS MORALLY SPECIAL?

There is an idea about how Darwinism might be related to ethics that is older and deeper than either 'evolutionary ethics' or sociobiology. Darwin's earliest readers realized that an evolutionary outlook might undermine the traditional doctrine of human dignity, a doctrine which is at the core of Western morals. Darwin himself seems to suggest this when he says that the conception of man as 'created from animals' contradicts the arrogant notion that we are a 'great work'. It is a disturbing idea, and Darwin's friends as well as his enemies were troubled by it. In explaining his initial reluctance to accept Darwin's theory, Lyell, for example, wrote:

You may well believe that it cost me a struggle to renounce my old creed. One of Darwin's reviewers put the alternative strongly by asking 'whether we are to believe that man is modified mud or modified monkey.' The mud is a great comedown from the 'archangel ruined.'

Surprisingly, philosophers have not taken this thought very seriously. I shall argue, however, that discrediting 'human dignity' is one of the most important implications of Darwinism, and that it has consequences that people have barely begun to appreciate.

Two Early Assessments of Our Problem

Huxley's lecture to the working men. In 1860 Thomas Henry Huxley delivered a series of lectures 'for working men only', to explain to them the shocking new idea that humans are descended from apelike ancestors. Huxley, then a 35-year-old professor at the Royal School of Mines, had been giving these 'people's lectures' (as he called them) for five years. He believed it was both possible and important for the working classes to understand the latest developments in science, and he meant to teach them: 'I am sick and tired of the *dilettante* middle class', he wrote, 'and mean to try what I can do with these hard-handed fellows who live among facts.' The 'people's lectures' were enormously popular; as many as 600 men might attend a single talk. We are told that a cabby once refused to accept payment from Huxley, saying 'Oh, no, Professor. I have had too much pleasure and profit from hearing you lecture to take any money from your pocket.' It is possibly an apocryphal story, but it accurately suggests the enthusiasm with which Huxley's lectures were received.

Huxley's commitment to these men was probably a consequence of his own humble beginnings. Unlike the other notable scientists with whom he would later be associated, Huxley was not well-to-do. As a boy he had only two years of formal schooling. A less gifted child might have drifted into a marginal life, but not Thomas Henry: he educated himself, reading science and logic and learning German on his own. By the time he was sixteen, he was apprenticed to a doctor, and at seventeen he won a Free Scholarship to study at the Charing Cross School of Medicine in London. At twenty he published his first scientific paper, on human hair, in the *Medical Gazette*.

In those days, service with the Royal Navy was a well-travelled route to scientific eminence. Voyages to the far parts of the world usually included scientific officers who would report on the geology, flora, and fauna of the lands visited. A young naturalist who had the opportunity to make such observations would be a leg up on his competitors. Darwin and Hooker had both become prominent after such voyages. Therefore it was natural for Huxley, who wanted to follow in their footsteps, to apply for a naval post. After completing his medical studies, he was commissioned as assistant surgeon on the frigate H M S *Rattlesnake*, and went off to Australia. The trip took four years.

When he returned to England in 1850, Huxley was ready to bring out a book on the oceanic animals he had studied on the voyage. He expected the Admiralty to pay the costs of publication. When they would not, he was furious. He felt a promise had been broken. In the ensuing dispute, Huxley refused to obey orders and was kicked out of the Navy. Thus, during the early 1850s he was broke, without prospects, without a university background, and with no experience other than that of a naval assistant-surgeon. To make things worse, he longed to marry a woman he had met in Sydney. Without money or prospects, he could not send for her.

But Huxley had been noticed by England's men of science, and his reputation was growing. He had published several papers, some of them sent back to England during his voyage, and within a year of his return he was elected to the Royal Society. A year later he was awarded the Society's Royal Medal and was made a member of the Society's Council. Yet his lack of pedigree held him back, and not until 1855 could he obtain a post that would pay enough to permit him to marry. Then, thanks to his connections in London's scientific circles, he was made Professor of Natural History at the Royal School of Mines, a post he would hold for more than thirty years. That same year he and his bride-to-be were reunited, and they soon began the family that was to produce such distinguished Huxleys as Leonard, Julian, and Aldous.

Having come up the hard way, it is not surprising that Huxley was enthusiastic about lectures for 'working men' who, like himself, had not had much opportunity for formal schooling. What is surprising, however, is the content of those lectures. We might expect watered-down science, avoiding technicalities and concentrating on 'popular' topics. But not so. Huxley attracted hundreds of unlettered men to lectures that presented real science in an uncompromising manner. His gifts as a speaker were formidable.

What, exactly, did he talk about? The subjects on which Huxley lectured, between 1855 and 1859, were unremarkable. The same subjects might have been chosen by any lecturer of the time. But in 1859 Huxley's outlook was changed forever, by the publication of *The Origin of Species*. Huxley was an instant convert to Darwin's view, and he was not a man to keep his support quiet. He had known Darwin for some time—in the small world of British science, everyone knew everyone else—and he knew that Darwin was not temperamentally given to public disputation. Huxley promptly volunteered to stand in for him. Soon after reading the *Origin*, Huxley wrote to Darwin:

I trust you will not allow yourself to be in any way disgusted or annoyed by the considerable abuse and misrepresentation which, unless I am greatly mistaken, is in store for you. Depend upon it you have earned the lasting gratitude of all thoughtful men. And as to the curs which will bark and yelp, you must recollect that some of your friends, at any rate, are endowed with an amount of combativeness which (though you have often and justly rebuked it) may stand you in good stead. I am sharpening up my claws and beak in readiness.

Huxley was as good as his word. In the years ahead he rarely missed an opportunity to expound and defend Darwin's theory. Thus in 1860 we find him explaining to his working men that they are kin to the apes.

Most of the 1860 lectures were devoted to expounding the anatomical evidence for this kinship. The structural differences between man and the gorilla, Huxley argued, are much smaller than the differences between the gorilla and the monkey; thus, if we admit kinship between the gorilla and the monkey, how can we deny kinship between man and the gorilla? To buttress this argument he cites detail after detail concerning the anatomy of hands, feet, teeth, jaw, and brain.

Towards the end of the lectures, however, Huxley turns to the question that must have been on all their minds: If we are only advanced apes, what of the dignity and worth of man? We think ourselves not only different from, but superior to, the other creatures that inhabit the earth. All our ethics and religion tell us this. Are we now to understand that we are no better than mere apes? Huxley himself puts it this way:

On all sides I shall hear the cry—'We are men and women, not a mere better sort of apes, a little longer in the leg, more compact in the foot, and bigger in the brain than your brutal Chimpanzees and Gorillas. The power of knowledge—the conscience of good and evil—the pitiful tenderness of human affections, raise us out of all real fellowship with the brutes, however closely they may seem to approximate us.'

Huxley, however, thought this worry is based on a misunderstanding, easily corrected. He asks, 'Could not a sensible child confute, by obvious arguments, the shallow rhetoricians who would force this conclusion upon us?' He was eager to reassure his audience that Darwinism has no adverse implications for the idea of human dignity. Even if we accept the idea that we are kin to the apes, he said, we can go right on thinking of ourselves as superior to, and somehow set apart from, the rest of creation. Although we may resemble the apes, we are of a different order. Huxley continued,

I have endeavoured to show that no absolute structural line of demarcation, wider than that between the animals which immediately succeed us in the scale, can be drawn between the animal world and ourselves; and I may add the expression of my belief that the attempt to draw a psychical distinction is equally futile, and that even the highest faculties of feeling and of intellect begin to germinate in lower forms of life. At the same time, no one is more strongly convinced than I am of the vastness of the gulf between civilized man and the brutes; or is more certain that whether *from* them or not, he is assuredly not *of* them. No one is less disposed to think lightly of the present dignity, or despairingly of the future hopes, of the only consciously intelligent denizen of this world.

This is reassuring, but what reasoning lies behind it? Huxley tells us that 'there is no absolute structural line of demarcation' that separates us from the other animals, and he adds that this is true not only regarding our physical characteristics, but our 'highest faculties of feeling and intellect' as well. If this is so, where is our superiority? His answer is an old familiar one, that has appealed to philosophers in all ages: we can talk, while other animals cannot.

Our reverence for the nobility of manhood will not be lessened by the knowledge that Man is, in substance and structure, one with the brutes; for, he alone possesses the marvelous endowment of intelligible and rational speech, whereby, in the secular period of his existence, he has slowly accumulated and organised the experience which is almost wholly lost with the cessation of every individual life in other animals; so that, now, he stands raised upon it as on a mountain top, far above the level of his humble fellows, and transfigured from his grosser nature by reflecting, here and there, a ray from the infinite source of truth.

Thus, Huxley's message was clear: we are kin to the apes, but we need not worry. It makes no difference to our exalted view of ourselves, or to our dismissive view of them. We are still men, noble and fit for reverence; and the brutes are still the brutes, without the 'marvelous endowment' that makes us special.

For Huxley, the idea that evolutionary thinking undermines human dignity was just another club that might be used by Darwin's enemies to discredit the new theory. It was nothing but a potential objection to Darwinism, to be disarmed as quickly as possible. Huxley was a polymath who would go on to write many books, including lengthy works about philosophy and ethics, but it never seems to have occurred to him that challenging the traditional idea of human dignity might be a positive contribution the new outlook could make. That thought did, however, occur to Asa Gray.

Asa Gray's lecture to the divines. In 1880 another of Darwin's champions delivered a series of lectures. Asa Gray, America's foremost botanist, gave two lectures at the Theological School of Yale College. He had been invited there to discuss 'difficult and delicate matters' concerning the relation between science and religion. In particular, the theologians wanted to hear about Darwin's revolutionary ideas. Gray was the natural man to invite. Just as Huxley was Darwin's great advocate in Britain, Gray was Darwin's leading defender in America. And, what was also important on this occasion, Gray was a lifelong churchman whose commitment to religious ideals could not be doubted.

Gray was born in 1810—one year and some months after Darwin—to parents of modest means in upstate New York. Like Huxley, he started out as a physician, and then turned to science. At age 25 he published a book, *Elements of Botany*, which gained him a good reputation, and at 31 he was made Professor of Botany at Harvard.

Gray had corresponded with Darwin, beginning in the 1850s, and was one of the few people to whom Darwin confided his new ideas prior to their publication in the *Origin*. In 1856 Darwin had written Gray a long letter in which he summarized his views on natural selection. (A letter which, as we have seen, was used as proof of Darwin's priority over Wallace.) Darwin later confessed: 'I thought you would utterly despise me when I told you what views I had arrived at.' But despite his religious convictions, Gray did not despise Darwin or his views. He was sympathetic from the first and soon became an enthusiastic convert. His early notice of Darwin's theory enabled him to begin arguing for it immediately upon its publication, while others were still trying to decide how to react.

Standing before the theological faculty and students at Yale, Gray was 70 years old and approaching the end of a distinguished career. He began with some modest remarks about his limitations as a theologian, and then launched into a long review of the state of biology, emphasizing, of course, the transforming contributions of Darwinism. (Interestingly, the science in Gray's presentation to the theologians was a good bit less demanding than that in Huxley's talks to his working men.) He then took up the vital question of whether evolutionary thought is compatible with religious faith. Gray had spoken and written on this many times before, so his words came as no surprise: true religion, he said, has nothing to fear from Darwin, and in fact can learn much from the perspective Darwinism provides. Evolutionary thought describes how a portion of the created order operates; but it neither says nor implies anything about the author of creation, his purposes, or his plans. Religion is still needed to make our picture of the world complete.

And what of man? If we share a common nature with the brutes, are we not thereby lowered to their level? Like Huxley, Gray saves this issue until last. Then he quickly reassures his audience that man is indeed special, because only man has the power of abstract thought. Again like Huxley, he associates this with man's ability to talk:

A being who has the faculty—however bestowed—of reflective, abstract thought superadded to all lower psychical faculties, is thereby *per saltum* immeasurably exalted. This, and only this, brings with it language and all that comes from that wonderful instrument; it carries the germs of all invention and all improvement, all that man does and may do in his rule over Nature and his power of ideally soaring above it. So we may well deem this a special gift, the gift beyond recall, in which all hope is enshrined.

To his credit, Gray does not leave this thought dangling, but tries to connect it to his general evolutionary outlook. The mental powers of man have evolved from lesser psychical powers that are still found in the lower animals—and this process, he hints, may be none other than the evolution of the soul! Thus evolutionary thought is made consistent with the idea that we have souls and the brutes do not. And souls alone are worthy of immortality. Appropriating Darwin's favourite terminology of a 'struggle for life', he says:

May it not well be that the perfected soul alone survives the final struggle of life, and indeed 'then chiefly lives'—because in it all worths and ends inhere; because it only is worth immortality, because it alone carries in itself the promise and potentiality of eternal life!

Thus Gray's message was not unlike Huxley's: we can accept our kinship to other animals without abandoning our view of ourselves as set apart from them. We are 'immeasurably exalted' by our higher mental powers; we 'soar above' the rest of nature, carrying within ourselves 'all worths and ends'.

If Gray had stopped there, his view would have been unremarkable. It would have been the same as Huxley's, and in fact the same as that of almost every other important defender of evolution of his day. But Gray added an extra, significant thought. Why, he wondered, do people *care* so much about this issue? Why are people so resistant to the idea that they are kin to the lower animals? Gray speculates that we resist evolutionary ideas because of their implications for morality—in particular, because of their implications for the morality of how we treat animals. If we were to acknowledge that we are kin to the animals, then it would be difficult to deny that, in so far as they are similar to us, they have the same rights that we have. But we do not wish to acknowledge this; we do not wish to give mere animals a moral claim on us. Therefore Gray, a humane man, found a 'meanness' in our wish to divorce ourselves from the rest of creation:

Man, while on the one side a wholly exceptional being, is on the other an object of natural history—a part of the animal kingdom . . . [H]e is as certainly and completely an animal as he is certainly something more. We are sharers not only of animal but of vegetable life, sharers with the higher brute animals in common instincts and feelings and affections. It seems to me that there is a sort of meanness in the wish to ignore the tie. I fancy that human beings may be more humane when they realize that, as their dependent associates live a life in which man has a share, so they have rights which man is bound to respect.

Is this additional thought consistent with Gray's earlier assertion that humans are 'immeasurably exalted' above the rest of nature? It is easy to see that competing ideas are struggling here. On the one hand Gray, like Huxley, wants to regard man as special. This means, among other things, that man has rights the other animals do not share. On the other hand, unlike Huxley, Gray was sensitive to the fact that evolutionary thinking makes this notion problematical. If humans resemble other animals so closely, mustn't we at least consider the possibility that, if man has rights, so do they? Here is, if not a problem, at least the recognition of an area where problems might arise.

The Idea of Human Dignity

What, exactly, is the traditional idea of human dignity? I do not mean to be asking about some esoteric doctrine advanced by a philosopher; instead, like Huxley, I am interested in the basic idea that forms the core of Western morals, and that is expressed, not only in philosophical writing, but in literature, religion, and in the common moral consciousness. This core idea has two parts, and involves a sharp contrast between human life and non-human life. The first part is that human life is regarded as sacred, or at least as having a special importance; and so, it is said, the central concern of our morality must be the protection and care of human beings. The second part says that non-human life does not have the same degree of moral protection. Indeed, on some traditional ways of thinking, non-human animals have no moral standing at all. Therefore, we may use them as we see fit.

This idea has a long history, and much of that history is intertwined with the history of religion. The great religions provide large-scale explanations of the nature of the world, its cause, and its purpose. Those explanations are almost always flattering to humans, assigning them a privileged place in the scheme of things. The idea that human beings have a special place in creation is so prominent, in so many religious traditions, that religion itself has sometimes been explained as an expression of man's desire to affirm his own worth.

The Western religious tradition, a blend of Judaism and Christianity, is a case in point. Man, it is said, was made in the image of God, with the world intended to be his habitation, and everything else in it given for his enjoyment and use. This makes man, apart from God himself, the leading character in the whole cosmic drama. But that is only the beginning of the story. Other details reinforce the initial thought. Throughout human history, God has continued to watch over and interact with man, communicating with him through the saints and prophets. One of the things he has communicated is a set of instructions telling us how we are to live; and almost all those instructions concern how we must treat other humans. Our fellow humans are not to be killed, lied to, or otherwise mistreated. Their lives are sacred. Their needs are always to be taken into account, their rights always respected. The concern we are to show one another is, however, only a dim reflection of the love that God himself has for mankind: so great is God's love that he even became a man, and died sacrificially to redeem sinful mankind. And finally, we are told that after we die, we may be united with God to live forever. What is said about the other animals is strikingly different. They were given by God for man's use, to be worked, killed, and eaten at man's pleasure. Like the rest of creation, they exist for man's benefit.

The central idea of our moral tradition springs directly from this remarkable story. The story embodies a doctrine of the specialness of man and a matching ethical precept. Man is special because he alone is made in the image of God, and above all other creatures he is the object of God's love and attention; the other creatures, which were not made in God's image, were given for man's use. We might call this the 'image of God thesis'. The matching moral idea, which following tradition we will call 'human dignity', is that human life is sacred, and the central concern of our morality must be the protection and care of human beings, whereas we may use the other creatures as we see fit.

Of course, many people do not believe the religious story, and consider their own thoughts about ethics to be independent of it. Yet a religious tradition can influence the whole shape of a culture, and even determine the form that secular thought takes within it. Only a little reflection is needed to see that secular moral thought within the Western tradition follows the pattern set by these religious teachings.

Few Western moralists have been satisfied to leave the idea of man's specialness stated in an overtly theological way. If we are made in the image of God, they reasoned, it should be possible to identify the divine element in our make-up. In what way, exactly, do we resemble the Almighty? The favoured answer, throughout Western history, has been that man alone is rational. Aristotle, expressing the Greek view of the matter, had said that man is the rational animal, and differs in this respect from all other creatures. This thought was put to use by the doctors of the Church: the divine element in man, they said, is his rationality. This we might call the 'rationality thesis': man is special because he alone is rational. Non-human animals are not rational, and so are not to be compared, in this regard, with humans.

In this way the doctrine of man's specialness was secularized, and cast into a form palatable even to those who are sceptical of the story behind the religious version of the idea. St Thomas Aquinas summarized all this—the rationality thesis, its relation to the image of God thesis, and their importance for the idea of human dignity—when he wrote that:

Of all parts of the universe, intellectual creatures hold the highest place, because they approach nearest to the divine likeness. Therefore the divine providence provides for the intellectual nature for its own sake, and for all others for its sake.

The idea of a unique human mental capacity—a capacity unlike anything to be found elsewhere in nature—may therefore be viewed as the secular equivalent of the idea that man was created in the image of God. It does the same work in our moral system, namely, it buttresses the idea that, from a moral point of view, humans are special. This means that, even if the image of God thesis is rejected, the matching moral idea need not be abandoned. Secular thinkers who reject religion can continue to believe in human dignity, and can justify doing so by pointing to man's unique rationality.

Some Practical Implications of the Idea of Human Dignity

The idea of human dignity has numerous practical consequences, both for the treatment of human beings and for the treatment of non-human animals. Often, the idea has taken an extreme form, with human life taken to be inviolable while non-human life is held to be utterly inconsequential.

The sanctity of innocent human life. More precisely, in our moral tradition *innocent* human life is taken to be inviolable. Guilty persons—criminals, aggressors, and soldiers fighting unjust wars—are not given this protection, and in some circumstances they may justly be killed. The innocent, however, are surrounded by a wall of protection that cannot be breached for any reason whatever. Such practices as suicide, euthanasia, and infanticide are violations of innocent life, and so they are not permitted. The moral rule governing such actions is simple: they are absolutely forbidden.

Suicide will serve as a convenient example (although euthanasia or infanticide would do just as well). One might think that, since the suicide takes only his or her *own* life, the prohibition upon it would not be so strict as the prohibition upon killing others. Prior to the coming of Christianity, the philosophers of Greece and Rome took this attitude.

Although they condemned cowardly suicides, they thought it could be permissible in special circumstances. The Christians, however, took a sterner view. St Augustine, whose thought shaped much of our tradition, argued that 'Christians have no authority for committing suicide in any circumstances whatever.' His argument was based mainly on an appeal to authority. The sixth commandment says 'Thou shalt not kill'. Augustine pointed out that the commandment does not say 'Thou shalt not kill *thy neighbour*'; it says only 'Thou shalt not kill', period. Thus, he argued, the rule applies with equal force to killing oneself.

Augustine held that man's reason is 'the essence of his soul', and in this he laid the foundation for later thought on the subject. A rational being, later thinkers would insist, can never justify doing away with himself, for he must realize that his own value is too great to be destroyed. St Thomas Aquinas, who held that man's rationality is central to his nature, argued that suicide is absolutely opposed to that nature. Suicide, he said, is 'contrary to that charity whereby every man should love himself'.

If Augustine and Aquinas were the towering figures of the Middle Ages, then the greatest of the modern philosophers, many would say, was Immanuel Kant. If we turn to Kant on suicide, we find that his views are almost indistinguishable from those of Augustine and Aquinas. Kant placed even more weight on the thesis of man's unique rationality than did Aquinas; his whole moral system was based on it. According to his famous formula, the ultimate moral principle is that we should treat human beings as 'ends in themselves'. Humans, he said, have 'an intrinsic worth, i.e., *dignity*', which makes their value 'above all price'.

If human life has such extraordinary worth, then it is only to be expected that a man can never justify killing himself. Kant draws this conclusion. Like Augustine and Aquinas, he believed that suicide is never morally permissible. His argument relies heavily on comparisons of human life with animal life. People may offer various reasons to justify self-murder, he says, but these attempted justifications overlook the crucial point that 'Humanity is worthy of esteem'. To kill oneself is to regard one's life as something of such little value that it can be obliterated merely in order to escape troubles. In the case of mere animals, this might be true—we kill animals to put them out of misery, and that is permissible. However, we should not think that the same may be done for a man, because the value of a man's life is so much greater: 'If [a man] disposes over himself,' Kant says, 'he treats his value as that of a beast.' Again, 'The rule of morality does not admit of [suicide] under any condition because it degrades human nature below the level of animal nature and so destroys it.'

All this follows, Kant thought, from taking the idea of man as a rational being (and therefore, as an exalted being) seriously. One might think, then, that there is no need to invoke religious notions to clinch the argument—the secular version of man's specialness should do the job alone, unaided by religious conceptions. However, Kant saw the secular argument and the religious story as working hand in hand. To secure the conclusion, he added:

But as soon as we examine suicide from the standpoint of religion we immediately see it in its true light. We have been placed in this world under certain conditions and for specific purposes. But a suicide opposes the purpose of his Creator; he arrives in the other world as one who has deserted his post; he must be looked upon as a rebel against God.

It is clear, then, that in Kant's mind—as in the minds of many others—the idea of man as a rational being was still closely linked to the idea of man as made in God's image.

The lesser status of non-human animals. The doctrine of man's specialness serves to exalt man at the expense of the other creatures that inhabit the earth: we are morally special, and they are not. Because we have a different nature, we have a moral standing that they lack. Once again, virtually all the important figures in our tradition agree on this. Aquinas was careful to point out that, although man's rationality gives him a special status, other animals have a very different place in the natural order. 'Other creatures', he said, 'are for the sake of the intellectual creatures.' Therefore, 'It is not wrong for man to make use of them, either by killing or in any other way whatever.' But shouldn't we be kind to them out of simple charity? No, Aquinas says, because they are not rational:

The love of charity extends to none but God and our neighbour. But the word neighbour cannot be extended to irrational creatures, since they have no fellowship with man in the rational life. Therefore charity does not extend to irrational creatures.

Kant, again, says much the same thing. Lacking the all-important quality of rationality, non-human animals are entirely excluded from the sphere of moral concern. It is man who is an 'end in himself'. Other entities have value only as means, to serve that end. Thus for Kant, animals have the status of mere things, and we have no duties to them whatsoever: 'But so far as animals are concerned,' he says, 'we have no direct duties. Animals . . . are there merely as means to an end. That end is man.' By a 'direct duty' Kant meant a duty based on a concern for the animal's own welfare. We may indeed have duties that *involve* animals, but the reason behind these duties will always refer to a human interest, rather than to the animals' own interests. Kant adds that we should not torture animals pointlessly, but the reason, he insists, is only that 'He who is cruel to animals becomes hard also in his dealings with men.' We are not, morally speaking, required to do (or refrain from doing) anything at all for the animals' own sakes.

It may seem that, in citing Aquinas and Kant, I have chosen extreme examples. Surely, one might say, our tradition is more complicated than this, and includes thinkers whose views are less unsympathetic to the animals. There is some justice in this complaint. In the biblical sources we find not only the idea that man has dominion over nature but also the contrasting notion that all of creation is to be revered as God's handiwork. On this latter conception, man's duty is to be a good steward of nature, not its exploiter. Someone who wanted to oppose the cruel exploitation of animals might cite this idea for support. Moreover, St Francis, not St Thomas, could be taken as one's model. St Francis, who is said to have preached to the birds, is remembered principally for proclaiming that all living creatures were his brothers and sisters, and for his gentle wonder at all of nature.

Yet there can be little doubt which of these two approaches has dominated Western culture. St Francis is a striking figure precisely because the legends about him contrast so dramatically with more orthodox ways of thinking. But real dissenters are harder to find than one might think; even St Francis appears to be less of a maverick when we examine his views more closely. When he talked to the animals, he heard them replying: 'Every creature proclaims: "God made me for your sake, O man!"' He regarded the animals as his brothers and sisters, but he took the same attitude towards the sun, the moon, wind, and fire. All were part of a creation to be revered, but equally they were all intended for man's use. Moreover, his kindness to the animals apparently did not extend to refraining from eating them—he did not recommend vegetarianism to his followers. On balance, then, it seems that St Francis accepted most of what I have called the doctrine of human dignity, and the religious cosmology that supported it, even though he combined it with a more reverent attitude than most other Christian thinkers. And in any case, as the historian John Passmore remarks, 'Francis had little or no influence.'

How Darwinism might Undermine the Idea of Human Dignity

The foregoing exposition is familiar enough; yet philosophers are apt to be impatient with it because of a point we have already discussed. The image of God thesis and the rationality thesis are, speaking loosely, matters of (purported) fact. The matching moral idea is a normative view. What, exactly, is supposed to be the relation between them? It cannot be that the latter follows logically from the former, because, as Hume observed, normative conclusions cannot legitimately be derived from factual premises. It would seem that the fragment of traditional morality we have been discussing is based on just this error—or at least, that my reconstruction commits the error. First we 'proceed in the ordinary way of reasoning' and 'establish the being of a God', that God made us in his own image, that we are uniquely rational creatures, and so forth. These are matters of what (allegedly) *is* the case. But then we go on to conclude from this that the protection of human life *ought* to be the purpose of our morality—and here the mistake creeps in, for factual statements can never by themselves logically entail evaluations. Hume considers this point to be of the very last importance. 'This small attention', he says, 'would subvert all the vulgar systems of morality.' There is no doubt that Hume thought the ideas we have been considering form one such 'vulgar system'.

Max Black refers to the general logical point, that facts do not entail evaluations, as 'Hume's Guillotine'. Hume's Guillotine might also come into our discussion in another way—it seems to provide a quick and easy answer to the worry of Huxley's working men. If we accept a Darwinian view of human origins, must we therefore abandon the idea of human dignity? No, for the facts of evolution do not, by themselves, entail any moral conclusions. Darwin's theory, if it is correct, only tells us what *is* the case with respect to the evolution of species; and so, strictly speaking, no conclusion follows from it regarding any matter of value. It does not follow, merely because we are descended from apelike ancestors, that our lives are less important. When Huxley asked, 'Could not a sensible child confute, by obvious arguments, the shallow rhetoricians who would force this conclusion upon us?', he might well have had this sort of 'obvious argument' in mind.

The majority of twentieth-century moral philosophers would agree. Moral philosophers have been largely indifferent to Darwin, and fear of Hume's Guillotine has been largely responsible for that indifference. 'The facts of evolution do not entail any normative conclusions': most philosophers have assumed that, once this simple observation has been made, there is little more to be said.

Nevertheless, the nagging thought remains that Darwinism does have unsettling consequences. The philosopher's reassurance that there will be no problem if we only remember to distinguish 'ought' from 'is' seems altogether too quick and easy. I believe this feeling of discomfort is justified. Matters are more complicated than a simple reliance on Hume's Guillotine would suggest.

Matters are more complicated, first, because our beliefs are often tied together by connections other than strict logical entailment. One belief may provide *evidence* or *support* for another, without actually entailing it. As evidence accumulates, one's confidence in the belief may increase; and as evidence is called into question, one's confidence may diminish. This is a common pattern.

I want to highlight a part of this process that we may call *undermining* a belief. The basic idea is that a belief is undermined by new information when the new information *takes away the support* of the belief. Suppose, for example, you believe that 'Hound Dog' was written by the great songwriting team of Jerry Lieber and Mike Stoller. You believe this because you read it in a copy of the *Elvis Newsletter*. But then you learn that the *Elvis Newsletter* is unreliable; it is produced by a careless fan who gets his facts wrong as often as right. So you come to doubt whether the newsletter can be trusted, and as a result, your confidence that Lieber and Stoller wrote 'Hound Dog' is weakened. You may even come to have no belief at all about the authorship of that song.

Notice what has happened. Your original reason for believing Lieber and Stoller wrote 'Hound Dog' was that the newsletter said so. But the fact that the newsletter said this does not *entail* that they wrote the song. Moreover, although you may stop believing they wrote it when you learn the newsletter is unreliable, the fact that the newsletter is un-reliable does not entail that they did not write it. We are not dealing with a series of logical entailments. Rather, we are dealing with a situation in which one believes something based on available evidence, and in which one modifies one's beliefs as new evidence appears.

We should be careful to distinguish between undermining a belief and merely decreasing someone's confidence in the belief. The latter is a psychological effect that can be brought about in any number of ways, including non-rational ways. The former, however, is a rational process. After you learn that the *Elvis Newsletter* is unreliable, it is *reasonable* for you to have less confidence that Lieber and Stoller wrote 'Hound Dog'. It is a matter of adjusting belief to evidence.

We should also emphasize the difference between undermining a belief and proving the belief to be false. Your new evidence does not prove that Lieber and Stoller did not write 'Hound Dog'. It merely takes away your reason for thinking that they did. The original belief could still be true—and in fact, 'Hound Dog' was written by Lieber and Stoller (not for Elvis but for Big Mama Thornton)—but, before it will be rational for you to believe it, you need another reason.

The situation is no different when evaluative judgements are involved. Suppose you are a member of the Songwriters Association and you say that Lieber and Stoller ought to be given the Association's Lifetime Achievement Award. You may legitimately be asked why, because any judgement about what ought to be done must have reasons in its support if it is to be taken seriously; otherwise, it can be dismissed as arbitrary or unfounded. So you say: Lieber and Stoller ought to be given this award because they wrote such classics as 'Stardust' and 'Hong Kong Blues'.

Is there anything wrong with your reasoning? That Lieber and Stoller wrote these songs is (if true) a matter of fact. That they ought to be given the Lifetime Achievement Award is an evaluation. Therefore, according to Hume's Guillotine, one does not 'follow from' the other. But what is the logical importance of this? Is it, as Hume said, 'of the last consequence'? Suppose someone who opposed a special award for Lieber and Stoller remembered Hume and objected that there is no logical entailment here. That would be true but irrelevant. In providing reasons, one need not be claiming that the facts logically entail the 'ought' judgement. Rather, one need only claim that they provide *good reasons* for accepting the judgement. That is a weaker, but still significant, claim.

This person might, of course, dispute your reasons in other ways. He might argue that 'Stardust' and Hong Kong Blues' are not good songs (although he would have little chance of winning *that* argument). Or, more simply, he could point out that in fact Lieber and Stoller did not write those songs; they were written by Hoagy Carmichael. This undermines your judgement by taking away its support. You do not, of course, have to abandon the idea that Lieber and Stoller should be honoured; but if you are to continue to maintain it, you must come up with some other reason in its support. Luckily, this is easy to do. Even though they did not write the two songs you mentioned, they did write any number of other classics: not only 'Hound Dog', but 'Love Potion No. 9', 'Kansas City', 'Yakety Yak', and many others.

Philosophers love artificial examples, such as the Lieber-and-Stoller example I have been using, because artificial examples are easily controlled. They can be kept simple and manageable and so can be used to illustrate logical points in an uncomplicated way. Examples drawn from real life are, in contrast, messy and confusing. They abound with inconvenient details that don't fit neatly into one's preconceived framework. Nevertheless, the process of undermining beliefs is an important part of reasoning in the real world as well as in philosophers' fantasies.

Consider, for example, the seventeenth-century debate about

embryological origins. In the seventeenth century many scientists believed in a view known as 'preformationism'—as the name suggests, they believed that each organism starts off with all its parts already formed. The development of the organism therefore consists merely in its growing bigger and bigger. As one writer put it, the embryo's development consists in 'a stretching or growth of parts'. Observations by scientists such as Marcello Malpighi (1628–94), who introduced the microscope into embryology, seemed to support preformationism. Peering through his primitive instrument, Malpighi thought he saw the 'rudiments' of the chicken already present in the hen's egg. Applied to human reproduction, preformationism became the theory of '*emboîtement*', which held that all future generations of humans were already present in Eve's ovaries—the emergence of new people being like the opening of an endless series of Chinese boxes.

Preformationists were also encouraged by another development in embryology, the theory of 'ovism'—the idea that all organisms develop from eggs. William Harvey (1578–1657), who, in addition to discovering the circulation of blood, was one of the first great embryologists, declared that 'An egg is the common origin of all animals.' 'Ovism' was a great advance over the received view that different types of creatures are generated in different ways—e.g. that plants and animals reproduce differently, that insects are generated spontaneously in rotten meat, and so on. It was a first attempt to provide a unified theory of generation. Ovism seemed to support preformationism, in that it permitted Malpighi's results to be generalized. If all organisms have a common origin (in eggs), and we can observe the 'rudiments' of chicks in chicken-eggs, then we might reasonably expect that similar rudiments will be present in other types of eggs.

Ovism, however, was a short-lived theory. In 1677 Leeuwenhoek first observed spermatozoa, and concluded that the female ovum alone could not be the sole source of the human embryo. Other evidence quickly mounted. A Swedish doctor, Niklaas Hartsoeker (1656–1725) calculated that if *emboîtement* were true, then the original egg in Eve's ovaries (in 4004 BC) would have to be larger than its present-day descendant by a factor of $10^{30,000}$—and that did not even allow for future generations. Ovism was therefore abandoned, and preformationism was deprived of an important bit of support.

But the abandonment of ovism did not lead straightaway to the rejection of preformationism. On the contrary, ovism was replaced by a new view, 'animaculism', which was interpreted in such a way as to support preformationism. Animaculism was a kind of reverse-image of ovism; it held that the egg is nothing but food for the developing organism, which is entirely contained in the male sperm. Despite his earlier calculation, Hartsoeker himself became an enthusiast for the new theory, and published a drawing of a fully formed infant curled up inside a spermatozoon. The demise of ovism was a setback for preformationists, but it was not until later that their view was finally discredited. What finally killed preformationism was, among other things, the simple realization that organisms inherit characteristics from both parents.

The story of the rise and fall of preformationism is not the story of scientist-logicians patiently drawing out the strict entailments of various discoveries. Malpighi's observations did not entail the truth of pre-formationism, nor did ovism: they were, however, evidence in its favour. They made preformationism plausible. Similarly, Hartsoeker's calculation was evidence against ovism, although it did not demonstrate incontrovertibly that ovism must be false. But by casting doubt on the truth of ovism, Leeuwenhoek's and Hartsoeker's results took away an important bit of support for preformationism. After ovism was rejected, it was reasonable to have less confidence in preformationism. It was a matter of adjusting belief to evidence. The original belief could still have been true—and in fact, it was defended by some scientists for many more years.

The debate over preformationism was also connected with moral matters in an instructive way. As we have seen, preformationism, with its accompanying doctrine of *emboîtement*, said that all the generations of humans were already present in Eve's ovaries, like dolls within dolls. This idea led the Christian Church to adopt a stricter view of abortion than it had previously accepted. St Thomas had held that the foetus does not become fully human until several weeks after conception, when it takes on a recognizably human form. Therefore, abortion in the early weeks of pregnancy had been tolerated. But *emboîtement* suggested that the foetus already has a human form—no matter how tiny it is—from the very beginning, and so abortion at any point was the killing of a real human being. The Church's moral stance was tightened accordingly, and abortion came to be condemned as morally wrong.

The Church had taken a moral position, and had given a reason in its support: if *emboîtement* is correct, then abortion is the killing of a tiny, perfectly formed human being. This is a matter of (alleged) fact. Now suppose a defender of abortion, remembering Hume, objected that the fact does not logically entail the evaluation. Again, that would be true but irrelevant. Any judgement about what ought to be done must have reasons in its support if it is to be taken seriously; otherwise, it can be dismissed as arbitrary or unfounded. But in providing reasons, one need not be claiming that the facts logically entail the moral judgement. One need only claim that they provide good reasons for accepting the judgement.

Eventually, of course, the theory of *emboîtement* was shown to be false, and now it is just a historical curiosity. Today we know that, in fact, foetuses develop from fertilized ova which are single cells and are nothing like fully formed humans, except that they contain human genetic material. This new information undermined the Church's position, by removing its support. Therefore, one might have anticipated that, after *emboîtement* was discredited, the Church would have returned to its earlier, more permissive attitude towards abortion. But it did not, and there was no reason why it had to, because the new information did not entail the falsity of the moral position. That position could still to be maintained, if other grounds could be found for it. And that is what happened: the Church found other arguments against abortion, and the stricter moral view was retained.

We can therefore see that Hume was both right and wrong. His point about the logical difference between factual and evaluative judgements may have been correct. But he was surely mistaken to think the point 'subverts all the vulgar systems of morality'—among which he included the moral system based on traditional religious ideas. Traditional morality is not subverted because in fact it never depended on taking the matching moral idea as a strict logical deduction from the image of God thesis or the rationality thesis.

What, then, is the relation between the image of God thesis (or the rationality thesis) and the matching moral idea? It is not that the former is supposed to entail the latter. Rather, it is that the former is supposed to provide good reason for accepting the latter. In traditional morality, the doctrine of human dignity is not an arbitrary principle that hangs in logical space with no support. It is grounded in certain (alleged) facts about human nature; those facts are what (allegedly) make it reasonable to believe in the moral doctrine. The claim implicit in traditional morality is that humans are morally special *because* they are made in the image of God, or because they are uniquely rational beings.

We are now in a position to explain how Darwinism might undermine traditional morality. The claim that Darwinism undermines traditional morality is not the claim that it entails that the doctrine of human dignity is false. It is, instead, the claim that Darwinism provides reason for doubting the truth of the considerations that support the doctrine. From a Darwinian perspective, both the image of God thesis and the rationality thesis are suspect. Moreover, there are good Darwinian reasons for thinking it unlikely that any other support for human dignity can be found. Thus, Darwinism furnishes the 'new information' that undermines human dignity by taking away its support.

This is how, despite Hume's Guillotine, Darwinism might undermine the traditional doctrine of human dignity. Does it actually do so? Before we can answer this question with confidence, several matters must be addressed. First, is Darwinism really inconsistent with a religious view of man's place in the world? Secondly, does an evolutionary perspective really cast doubt on man's status as a uniquely rational animal? And thirdly, even if the first two questions are answered affirmatively, and we have to give up the image of God thesis and the rationality thesis, might not there still be some other way of defending traditional morality, so that we can continue to accept something like the idea of human dignity? If not, what sort of morality should we accept? What sort of moral view is consistent with a Darwinian understanding of nature and man's place in it? In the following chapters we will take up these questions.

3 Must a Darwinian be Sceptical about Religion?

Kick out the image of God thesis

BISHOP WILBERFORCE believed that *The Origin of Species* was 'absolutely incompatible' not only with the history of origins given in the Bible, but with Christianity's 'whole representation of the moral and spiritual condition of man'. Like Wilberforce, many religious conservatives continue to condemn evolutionary ideas because they see them as contradicting revealed truth. In 1987 the largest Protestant denomination in the United States, the Southern Baptist Convention, became the latest in a long line of religious bodies to denounce Darwinism, when it officially adopted the view that Adam and Eve were real people and prohibited the teaching of anything else in its theological seminaries.

But religious people are not always hostile to evolution. Many take a more hopeful view. Liberal theologians have argued that the theory of evolution, like other scientific discoveries, only reveals in greater detail how God has chosen to order his creation. Natural selection, on this way of thinking, is just the means God has used to bring about the world that he wished to make. Henry Ward Beecher, the most prominent American preacher of the late nineteenth century, even suggested that an evolutionary perspective *adds to* the glory of God's creation; for, as he put it, 'Design by wholesale is grander than design by retail.' The Roman Catholic Church, after some waffling, seems to have settled on the compromise of St George Jackson Mivart: in 1953 Pope Pius XII proclaimed that 'The teaching of the Church leaves the doctrine of evolution an open question, as long as it confines its speculations to the development, from other living matter already in existence, of the human body'—however, he immediately adds, 'That souls are immediately created by God is a view which the Catholic faith imposes on us.'

If religious people are divided in their opinions about evolution, evolutionists are no less divided in their views about religion. Many evolutionary biologists are devout church-goers. Stephen Jay Gould cites this as evidence that 'the warfare between science and religion' is based on a misunderstanding. 'Most scientists', he writes, 'show no

99

hostility to religion. Why should we, since our subject doesn't intersect the concerns of theology?' Gould points out that Theodosius
Dobzhansky, 'the greatest evolutionist of our century', was a theist.
This alone, he says, should make us reject the notion of any intrinsic
incompatibility:

Unless at least half my colleagues are dunces, there can be—on the most raw and
direct empirical grounds—no conflict between science and religion. I know
hundreds of scientists who share a conviction about the fact of evolution, and
teach it in the same way. Among these people I note an entire spectrum of
religious attitudes—from devout daily prayer and worship to resolute atheism.
Either there's no correlation between religious belief and confidence in
evolution—or else half these people are fools.

Gould has a point, but not a very powerful one. If Darwinism and
theism were incompatible, it would not follow, as he suggests, that half
the evolutionary biologists are fools: it would only follow that they are
mistaken, and it would not be the first time that a large number of
intelligent people have been mistaken about something. Nevertheless,
the fact that so many thoughtful scientists are theists does provide *some*
evidence that evolution and theism are compatible—after all, if they
were incompatible, how could so many sensible people believe in both?
But it is inconclusive evidence at best.

We want to know more than what people happen to think. We want to
know whether evolution and religion *really are* compatible. Of course,
the answer depends on what religious view we consider. At one extreme,
religion might be conceived to be nothing more than some sort of
reverent attitude towards the world. Darwinism would certainly be
compatible with that. At the other extreme is fundamentalist Christianity, with its insistence on the literal truth of the creation story in
Genesis. Darwinian evolution is certainly not compatible with that. To
give our question a reasonable focus, however, we may settle on a view
somewhere in between. We may ask: Is Darwinian evolution compatible
with *theism*, broadly conceived? More specifically, remembering our
concern with the doctrine of human dignity, we will want to know the
answer to this slightly different question: Is Darwinian evolution compatible with a version of theism rich enough to support the 'image of
God thesis'—the idea that humans are made in God's image and enjoy a
special place in his creation?

DARWIN'S SCEPTICISM

As a preliminary to the main argument of this chapter, I want to say
something about Darwin's own attitude towards religion. As a young

man Darwin had planned to become a parson, and with a good conscience: for, as he said, 'I did not then in the least doubt the strict and literal truth of every word in the Bible.' This simple faith gradually left him, however, and by the time he was an old man he no longer believed in any religious doctrine. At age 67 he wrote, looking back, that 'Disbelief crept over me at a very slow rate, but was at last complete. The rate was so slow that I felt no distress, and have never since doubted even for a single second that my conclusion was correct.'

The erosion of belief was already noticeable by the time of his marriage. Emma was distressed by it, and wrote her new husband an eloquent letter urging him to reconsider the direction of his thinking. But even then it was too late. He had already been converted to evolutionism, and had discovered the theory of natural selection, and in his mind traditional religious belief was not compatible with his new outlook. Emma never gave up hoping that Charles would one day return to more conventional ways of thinking.

In his published works Darwin did not discuss religion directly, although he did drop a number of hints that his theory was not to be taken as compatible with traditional belief. His neglect of the subject was due largely to his desire not to complicate matters by taking on too many issues at once—it was quite enough to defend evolutionism, without attacking religion at the same time—and to his reluctance to cause his family needless distress. But there was still another reason. Darwin believed that direct attacks upon religion are not effective—people rarely change their religious views because of arguments. Patient scientific work, on the other hand, does have a long-term effect. As science progresses, the basis of religious belief is eroded, and such belief becomes harder and harder to maintain. (Perhaps Darwin was thinking of his own experience here.) In a letter written in 1880, two years before he died, he explained that:

Though I am a strong advocate for free thought on all subjects, yet it appears to me (whether rightly or wrongly) that direct arguments against Christianity and Theism produce hardly any effect on the public; and freedom of thought is best promoted by the gradual illumination of men's minds, which follows from the advancement of science. It has, therefore, been always my object to avoid writing on religion, and I have confined myself to science.

It is, therefore, only in the writings Darwin did not intend for publication—especially his letters and the *Autobiography*—that we find clear statements of his views concerning religion.

The *Autobiography* was written six years before Darwin died, and was intended for the private reading of his children and grandchildren. It was published a few years after his death, however, when his son Francis brought out the volume of 'Life and Letters' which was the usual memorial to great men in the nineteenth century. The *Autobiography* included an 11-page essay on religion. But when it was published, the most tartly anti-religious passages were omitted, at the request of Mrs Darwin, who said that they were too crude and insufficiently thought-out to represent her husband's considered judgements. Even the edited version aroused strong feelings within the family. Darwin's daughter Henrietta spoke of taking legal steps to halt its publication, although she never actually did so. The family did not make the unexpurgated *Autobiography* available until 1958, when a granddaughter, Nora Barlow, finally published the complete text.

Mrs Darwin's objections were caused, no doubt, more by her own distress over her husband's lack of faith than by any actual crudity of expression. Darwin's statement of his views in the *Autobiography* has the same crispness and elegance that we find in all his writings. And his discussion, though short, is thorough: he covers almost all the basic arguments that sceptically inclined thinkers have considered important. Some of these arguments are familiar and Darwin adds little new to them. But to others he gives a distinctive twist by connecting them with natural selection.

Darwin's attack on religion is directed partly at Christianity. When people accept religious belief, it is some particular religion that is accepted, and it is no less true that when sceptics reject religion, they are rejecting some particular religion—usually, the dominant religion of their culture. Thus, in our society, when people decide to throw off religion, it is Christianity they are discarding; and they feel compelled to give arguments against *it*, and not against Hinduism or Buddhism, even though they do not believe in Hinduism or Buddhism either. Darwin was no exception.

Darwin begins his discussion of religion in the *Autobiography* by arguing that we have no good reason to believe Christianity is true. The Bible cannot be trusted, he says, because it contradicts itself on important points, and its reports of miracles are not credible to those familiar with the lawlike workings of nature. What is more, the Hindus and Buddhists also have their sacred writings—why should we think the Christian writings are any more reliable than the others?

If the Bible cannot be trusted, what else could provide evidence that Christianity is true? One suggestion might be that religious faith is founded upon the feelings and experiences of believers. History is full of reports of special revelations and mystical experiences. Even ordinary people, who claim no special revelation, often say that they experience a sense of absolute certainty where religion is concerned, which no argument could shake. Thus religious conviction often has a kind of immediacy which seems to render further evidence unnecessary.

Darwin dismisses this line of thought with a few brief comments. Of course, he says, Christians have 'inward feelings' that support their views; that is only to be expected, since they have been raised in Christian cultures. But these feelings are worth nothing as evidence, for Buddhists, Mohammedans, and even barbarians have similar feelings concerning the truth of *their* convictions. Darwin says, 'This argument would be a valid one if all men of all races had the same inward conviction of the existence of one God; but we know that this is very far from the case.' Anyway, he says, it is easy to explain the existence of religious feelings without assuming they are intimations of truth. They might easily be a kind of wishful thinking: we do not want to die, and so we believe in a God who will see to it that we live forever.

But Darwin's rejection of Christianity is not based merely on our lack of evidence for its truth. He finds its doctrines unacceptable on independent grounds. The doctrine of hell, for example, he thinks should be rejected on moral grounds alone:

I can indeed hardly see how anyone ought to wish Christianity to be true; for if so the plain language of the text seems to show that the men who do not believe, and this would include my Father, Brother and almost all my best friends, will be everlastingly punished.
And this is a damnable doctrine.

Orthodox Christianity, therefore, is unacceptable.

The larger issue, however, is theism itself. Darwin was well aware that religion could be separated from uncritical dependence on specific texts, and that it could be purged of such 'damnable doctrines' as hell-for-the-unbelievers. Therefore, a more enlightened theism would be unaffected by these objections. Recognizing this, Darwin went on to give reasons for rejecting *any* broadly theistic belief. In this he relied, first of all, on the argument from evil.

The Problem of Evil

The existence of evil has always been a chief obstacle to belief in an all-good, all-powerful God. How can God and evil co-exist? If God is perfectly good, he would not want evil to exist; and if he is all-powerful, he is able to eliminate it. Yet evil exists. Therefore, the argument goes, God must not exist.

Expressed in such a simple way, the argument is easy to answer. One need only observe that God might have a good reason for permitting evil to exist. Through the centuries, theologians have suggested various possibilities:

1. Perhaps evil is necessary so that we may better appreciate the good. Indeed, we could not even know what goodness *is* if we did not have evil for comparison.
2. Perhaps evil is punishment for man's sin. Before the Fall, people lived in Paradise. It was their own sin that resulted in their expulsion. Therefore, people suffer because they have brought it on themselves.
3. Perhaps evil is placed in the world so that, by struggling with it, human beings can develop moral character. By striving to cope with adversity we develop such qualities as courage, compassion, and perseverence. If we lived in a perfect environment we would be slugs.
4. Perhaps evil is an unavoidable consequence of man's free will. In order to make us moral agents, rather than mere robots, it was necessary for God to endow us with free will. But in making us free agents, God enabled *us* to cause evil, even though he would not cause it himself.
5. Or, if all else fails, the theist can always fall back on the idea that our limited human intelligence is insufficient to comprehend God's great design. There is a reason for evil; we just aren't smart enough to figure out what it is.

All these arguments are available to reconcile God's existence with evil. Certainly, then, the simple version of the argument from evil does not *force* the theist to abandon belief. There are too many possible rejoinders. Therefore, any sceptic who advances this argument needs to explain why he thinks such rejoinders are insufficient.

Sceptics have traditionally offered two such explanations. First, they emphasize that an *excessive amount* of evil exists in the world. The first three theistic rejoinders might suffice to justify some smaller amount of evil, but not the great amount we actually find:

1. Even if some evil is necessary for us to appreciate the good, it is not necessary that there be so *much*. If, say, only half the number of people died every year of cancer, that would be plenty to motivate the appreciation of health.
2. If only the wicked suffered, it might make sense to think that suffering is punishment for sin. But good people suffer also. The theist might reply that all have sinned, and so all deserve punishment. But what about innocent babies, who sometimes have terrible diseases and die horribly? Of course the notorious doctrine of Original Sin

says that even babies are sinners. If this is supposed to mean that a new-born baby deserves to have epidermolysis bullosa—a virtually untreatable disorder characterized by widespread and constant blistering of the skin, so that there is no part of the body on which the infant can lie without pain—then surely this is one of those morally damnable doctrines of which Darwin complains.

3. Even if we develop moral character by confronting evil, there is no need for us to be overwhelmed by it. The amount of evil in the world could be reduced by two-thirds, and there would still be more than we could cope with.

Therefore, even if God found it necessary to permit the existence of some evil to accomplish these purposes, he still would not find it necessary to permit so much.

Secondly, sceptics have emphasized a distinction between *moral evil*, which is caused by human beings, and *natural evil*, which exists independently of human action. This distinction undermines theistic rejoinder number (4). Certainly some of the world's evil is the result of human choices—human beings choose to wage war and murder and rape and so forth. However, much suffering is also caused by forces over which humans have no control. Epidermolysis bullosa is one example, but there are many others: a volcano unexpectedly erupts and spills burning lava onto a village; or a tidal wave inundates a coastal town. The appeal to God's wish to allow humans freedom of choice cannot explain why he permits such 'natural' evils to exist.

Like many doubters before him, Darwin appealed to the existence of evil as a reason for rejecting theism. But he strengthened the argument by giving it two distinctive twists. First, Darwin, realized that the traditional debate centres entirely on evils related to *human* life and history. The evils that need justifying have to do with *man's* suffering, and the traditional theistic justifications have to do with *man's* comprehension of goodness, *man's* sin, *man's* moral development, and *man's* free will. But, typically, the first thing that occurs to Darwin is that human life and history are only a small part of nature and its history. Countless animals have suffered terribly in the millions of years that preceded the emergence of man, and the traditional theistic rejoinders do not even come close to justifying *that* evil. Darwin writes:

That there is much suffering in the world no one disputes. Some have attempted to explain this in reference to man by imagining that it serves for his moral improvement. But the number of men in the world is as nothing compared with that of all other sentient beings, and these often suffer greatly without any moral improvement. A being so powerful and so full of knowledge as God who could create the universe, is to our finite minds omnipotent and omniscient, and it *how could God let this go on to animals*

revolts our understanding to suppose that his benevolence is not unbounded, for what advantage can there be in the sufferings of millions of the lower animals throughout almost endless time?

The standard formulation of the problem, and the standard theistic rejoinders, subtly assume that humans have always been around. When it was thought that man has existed from the beginning, and that the natural world has always been pretty much as we know it, it may have been plausible to think of evil as a condition necessary for our improvement, our free will, and the like. But an evolutionary perspective puts the problem in a new and more difficult form.

Darwin also connected the argument to natural selection in a second and deeper way. In his view, divine creation and natural selection are rival hypotheses, alternative ways of explaining why nature is as it is. The argument from evil points out that divine creation is a poor hypothesis because it fits the facts badly. On this hypothesis, not only would we not *expect* evil to exist, it is impossible to explain it even after the fact. That is why, to save the hypothesis, its defenders are driven to say (5) that our limited human intelligence is insufficient to understand God's great design—saying which is no less than an admission that, on the hypothesis of divine creation, the problem of evil is insoluble.

But what of the alternative hypothesis, natural selection? What would it lead us to expect? Darwin contends that it fits the facts very well. 'The presence of much suffering', he says, 'agrees well with the view that all organic beings have been developed through variation and natural selection.' Why? In order to survive, an animal must be motivated to act in ways conducive to its self-preservation. Pain and pleasure are the motivators. In the absence of food, we suffer hunger; and when we eat, we enjoy it: together, these ensure that we eat. When enemies are nearby, we suffer fear; when we flee, we feel relieved; together, these ensure that we keep safe. Actually, Darwin thinks, pleasure is the more common motivational force, and that is why, in his view, the world contains more happiness than misery:

But pain or suffering, if long continued, causes depression and lessens the power of action; yet is well adapted to make a creature guard itself against any great or sudden evil. Hence it has come to pass that most or all sentient beings have been developed in such a manner through natural selection, that pleasurable sensations serve as their habitual guides.

Thus, Darwin believed, natural selection accounts for the facts regarding happiness and unhappiness in the world, whereas the rival hypothesis of divine creation does not.

The Ultimate Origin of the Universe

In Darwin's view one of the strongest arguments in favour of religious belief was the 'first cause' argument, and at times it tempted him to accept theism. This argument begins by asking why the universe exists at all—what brought the whole thing into being? It is then suggested that we cannot explain this without resorting to some sort of theological account. Thus, it is said, belief in God is justified, at a minimum, as the belief in a 'first cause' of the universe.

This is the most abstract and general of the theistic arguments. It does not suppose that God has any specific character beyond that of being the creator: it does not suppose that he is the Jehovah of the Jews or the Allah of the Moslems; it does not suppose that he is a 'revengeful tyrant', or that he loves mankind. It *does* rest on the belief that everything must have a cause, and that the chain of causes cannot stretch back indefinitely. We must at some point come to a first cause, and, as Aquinas put it, 'that cause we call God'.

For most of his life Darwin was impressed by this argument, and at times, he said, he was persuaded by it. Even after he had abandoned traditional Christianity, he sometimes thought that the hypothesis of God might play at least one legitimate part in a rational system of belief: at the very least, it might be used to explain the ultimate origin of the universe, which was otherwise inexplicable. He continued to believe this, off and on, as late as the publication of the *Origin*. But on the whole his attitude towards this hypothesis seems to have been one of distrust. Ultimately he rejected it, but because his rejection was tentative and undogmatic, some commentators have viewed Darwin as at least a qualified theist. In 1860, a few months after publication of the *Origin*, he wrote to Asa Gray, a staunch believer:

I had no intention to write atheistically. But I own that I cannot see as plainly as others do, and as I should wish to do, evidence of design and beneficence on all sides of us. There seems to me too much misery in the world. I cannot persuade myself that a beneficent and omnipotent God would have designedly created the Ichneumonidae with the express intention of their feeding within the living bodies of Caterpillars, or that a cat should play with mice. Not believing this, I see no necessity in the belief that the eye was expressly designed. On the other hand, I cannot anyhow be contented to view this wonderful universe, and especially the nature of man, and to conclude that everything is the result of brute force. I am inclined to look at everything as resulting from designed laws, with the details, whether good or bad, left to the working out of what we may call chance. Not that this notion *at all* satisfies me. I feel most deeply that the whole subject is too profound for the human intellect. A dog might as well speculate on the mind of Newton. Let each man hope and believe what he can. Certainly I agree with you that my views are not all necessarily atheistical.

Several points are plain. Even as he published the *Origin*, Darwin did not want to be called an atheist. When he contemplated the 'wonderful universe', he was *tempted* to conclude that it is the product of divine creation. But he refused to draw that conclusion himself, even though he could understand why others would do so. He distrusted the power of the human intellect to deal with the question of ultimate origins.

This became pretty much Darwin's standard line. When he was asked his view of religion, he would repeat these points again and again. These same themes recur, for example, in a letter to a Dutch student written thirteen years later:

I may say that the impossibility of conceiving that this grand and wondrous universe, with our conscious selves, arose through chance, seems to me the chief argument for the existence of God; but whether this is an argument of real value, I have never been able to decide. I am aware that if we admit a first cause, the mind still craves to know whence it came, and how it arose ... The safest conclusion seems to me that the whole subject is beyond the scope of man's intellect; but man can do his duty.

Here, incidentally, Darwin alludes to the standard traditional objection to the first-cause argument. The argument is motivated by the thought that nothing can exist without a cause, so we posit God as the cause of the universe. But this only invites the further question: What caused God? And if we are willing to think of God himself as uncaused, why not think of the universe as uncaused? Thus the argument fails, even on its own terms: the craving that motivates the argument (the craving to have everything causally explained) cannot be satisfied even if the argument is accepted (because we are still stuck with at least one thing, God, that is left uncaused).

Darwin's statements in his letters, although they leave little doubt about his fundamental scepticism, seem a bit wishy-washy in comparison to the robust proclamations of the *Autobiography*. Remember that in the *Autobiography* he declared that: 'Disbelief crept over me at a very slow rate, but was at last complete. The rate was so slow that I felt no distress, *and have never since doubted even for a single second that my conclusion was correct.*' Can this ringing declaration be squared with Darwin's other statements? Can it be reconciled with his having told Asa Gray that he could not be content to view the 'wonderful universe' as the result of 'brute force', or with his having told the Dutch student that he had 'never been able to decide' whether the first-cause argument is correct?

One view, suggested by Neal C. Gillespie, is that the more strident statements of the *Autobiography* were directed at the particular doctrines of Christianity, which Darwin no doubt rejected. However, on the larger issue of theism itself, Darwin was more open-minded. In particular, Darwin left open the possibility that the first-cause argument might be correct. Thus Gillespie attributes to Darwin at least a tenuous theism, and he argues that those who portray Darwin as a straightout unbeliever are taking too simple a view.

There is, however, another, more natural understanding of the matter. Francis Darwin explains the undogmatic tone of his father's letters by remarking that 'He naturally shrank from wounding the sensibilities of others in religious matters.' Such reticence would be easy to understand. We are all familiar with the awkward position of nonbelievers who live among family and friends who do not share their doubts. Some unbelievers are aggressive in their unbelief and take delight in picking at, or even ridiculing, the unsophisticated faith of their neighbours. Darwin was not like that. He loved his family and friends and did not enjoy coming into conflict with them. On the contrary, his differences with them made him miserable. As I noted above, Emma wrote to him shortly after their marriage, urging him to reconsider his rejection of orthodox religion. Darwin preserved the letter with his private papers and wrote at the bottom: 'When I am dead, know how many times I have kissed and cried over this.'

It is easy to understand, then, why Darwin would bend over backwards to avoid playing the village atheist. Indeed, the very word 'atheist' has harsh, dogmatic connotations: it suggests that one claims to *know* that theism is false. Especially in his letters, Darwin's approach was always to avoid dogmatism and to allow, in so far as he could honestly do so, that others might be right, or at least that they might have some good reasons for their beliefs. The first cause argument seemed to him the best reason, so in the letters he did not absolutely reject it, although clearly he did not consider it to be convincing. In the *Autobiography*, however, he was less reticent: 'Can the mind of man, which has, as I fully believe, been developed from a mind as low as that possessed by the lowest animal, be trusted when it draws such grand conclusions?'

Calling oneself an atheist, while surrounded by loved ones pained by one's doubts, is a hard thing. And so the distinction between atheism and agnosticism was useful to him, as it has been to so many in that position. The word 'agnostic' was new; it had recently been coined by Huxley, who described its genesis in his usual charming way:

When I reached intellectual maturity and began to ask myself whether I was an atheist, a theist, or a pantheist; a materialist or an idealist; a Christian or a freethinker; I found that the more I learned and reflected, the less ready was the answer; until, at last, I came to the conclusion that I had neither art nor part with any of these denominations, except the last. The one thing in which most of these good people were agreed was the one thing in which I differed from them. They were quite sure they had attained a certain 'gnosis'—had, more or less successfully, solved the problem of existence; while I was quite sure I had not, and had a pretty strong conviction that the problem was insoluble . . .

This was my situation when I had the good fortune to find a place among the members of that remarkable fraternity of antagonists, long since deceased, but of green and pious memory, the Metaphysical Society. Every variety of philosophical and theological opinion was there, and expressed itself with entire openness; most of my colleagues were -*ists* of one sort or another; and, however kind and friendly they might be, I, the man without a rag of a label to cover himself with, could not fail to have some of the uneasy feelings which must have beset the historical fox when, after leaving the trap in which his tail remained, he presented himself to his normally elongated companions. So I took thought, and invented what I conceived to be the appropriate title of 'agnostic'. It came into my head as suggestively antithetic to the 'gnostic' of Church history, who professed to know so much about the very things of which I was ignorant; and I took the earliest opportunity of parading it at our Society, to show that I, too, had a tail, like the other foxes.

And near the end of his life Darwin wrote to a man named Fordyce:

In my most extreme fluctuations I have never been an Atheist in the sense of denying the existence of a God. I think that generally (and more and more as I grow older), but not always, that an Agnostic would be the more correct description of my state and mind.

Agnostics, like atheists, are people who do not believe. But they wish to remain undogmatic about it.

Darwin's great contribution to the debate about religion was not, however, his discussion of the argument from evil, or the first-cause argument, or any of the other arguments that have been mentioned thus far. His great contribution was the final demolition of the idea that nature is the product of intelligent design. When we turn to this aspect of Darwin's thinking, it becomes clear why he believed that 'the gradual illumination of men's minds, from the advancement of science' leads to the abandonment of theism.

NATURE WITHOUT PURPOSE

When Karl Marx first read the *Origin*, he recognized its most revolutionary feature at once: 'Not only', he said, 'is a death blow dealt here for the first time to "Teleology" in the natural sciences but their rational

don't need a creator if there is natural selection

purp

meaning is empirically explained.' Marx had identified the philosophical nerve of the theory. To be properly understood, Darwin's rejection of the design hypothesis must be seen in the broader context of his overall rejection of teleology in nature.

A teleological explanation is an explanation of something in terms of its function or purpose: the heart is for pumping blood, the lungs are for breathing, and so on. Teleological explanations have always been thought to be indispensable in biology. Indeed, it is hard to see how they could be eliminated. The concept of the organism as a functional system is basic to biology. If we say only that the heart is a large muscle through which blood passes, we have omitted crucial information. We need to mention its relation to the other parts of the organism. Blood carries oxygen and nutrients without which the organism would die; but without the heart, the blood would not circulate—hence, the heart's 'purpose'. If we do not understand this, we have not understood the heart.

Yet when we say or imply that the heart has this purpose, we immediately encounter difficulties. It is easy enough to see that an artefact, such as a knife, has a purpose, because it is consciously made for a purpose. People make knives to use in cutting. This determines what the knife is for. If it later turns out that the knife can be used in some other way—say, as a screwdriver—that does not mean its purpose is to drive screws. The purpose is determined by the intentions of the maker. But if there is no maker—if the object in question is not an artefact—does it make sense to speak of a 'purpose'? If we find a piece of sharp rock, and use it to cut, can we therefore say that this is its purpose? No; we can only say that *our* purpose is to cut, and that we are using the rock for our purpose. *It* does not have a purpose. Purposes seem to depend on conscious intentions; otherwise, talk of purposes seems arbitrary and anthropomorphic. Another way to point up the difficulty is this: it is a fact that the heart pumps blood, but it is also a fact that it makes heart-sounds. What grounds are there for saying that the former, but not the latter, is its purpose? Why not say that its purpose is to make sounds?

The traditional solution was to accept the connection between purpose and conscious intention, and to say that biological structures have purposes precisely because they *are* designed, by God, who has their functions in mind as he creates them. Pre-Darwinian biologists thought this perfectly reasonable. After all, the hypothesis of divine creation was already an established part of their overall view. It was needed, they thought, to account for the beautiful, complex adaptations of the biological world. Since God's creative activity was already being assumed, why not make the most of it, and use it as a basis for teleological explanations as well?

It was a convenient, satisfying solution to the problem. But after Darwin this solution was no longer to be available. He demonstrated that even the most intricate adaptations could be accounted for without assuming any conscious design; all that was needed was random variation and natural selection. Biological structures are what they are, not because parts have been designed to 'fit' with the whole, but because variations have conferred advantages in the struggle for life. The whole organism is just the evolutionary sum of these variations. We may speak of an organ's 'function' because of the part it plays in enabling the organism to survive, but that is all. The connection between function and conscious intention is, in Darwin's theory, completely severed.

It is an exaggeration to say that Darwin dealt teleology a 'death blow'; even after Darwin we still find biologists offering teleological explanations. But now they are offered in a different spirit. Biological function is no longer compared to the function of consciously designed artefacts; and the fact—if it is a fact—that biological structures have purposes is no longer taken as an indication of how they were 'meant' to be. Thus the distinguished philosopher of science Ernest Nagel wrote in 1961,

It is a mistaken supposition that teleological explanations are intelligible only if the things and activities so explained are conscious agents or the products of such agents. Thus, in the functional explanation of lungs, no assumption is made, either explicitly or tacitly, that the lungs have any conscious end-in-view or that they have been devised by any agent for a definite purpose.

This is the standard post-Darwinian view. Biologists before Darwin would not have thought this disclaimer either necessary or appropriate.

Aristotle and Galileo

Darwin's theory brought to an end a way of understanding nature that had prevailed for many centuries. The Greeks believed that everything in nature exists for a reason or purpose. Aristotle, whose influence on the course of Western science was immense (and some say perverse) taught that in order to understand anything four questions must be answered: First, what is it? Secondly, what is it made of? Thirdly, how did it come to exist? And finally, what is it for? This last question, like the others, could be asked of anything whatever: his assumption was that everything has a purpose.

For Aristotle, there were no important differences of principle between the biological and physical sciences. Indeed, he seems to have taken biology as the paradigm, assuming that whatever is true of the biological world must be true of the physical world also. If biological structures exhibit purpose, he thought, physical structures do so as well.

'Nature', he said, 'belongs to the class of causes which act for the sake of something.' The rain falls, not 'of necessity', but in order to make the plants grow. Given this way of thinking, it is not surprising that Aristotelian physics appealed to all sorts of evaluative concepts in explaining the world. Gravitational attraction—or, what we would call gravitational attraction—was explained, for example, in terms of objects 'seeking their proper places'.

The rise of Christianity only strengthened the hold of Aristotelian conceptions. His science was found by the Church to be perfectly congenial. If nature manifests value and purpose, this could be seen easily enough as God's value and purpose. After all, according to the Christians, God had designed and created the world and everything in it—so it was only to be expected that everything that happens will happen for a purpose. The rain *does* fall so that the plants can grow. And if, in addition, the best available scientific theory of the universe places the earth at its centre, so much the better, for this symbolizes the importance of the earth in God's plan.

This way of thinking prevailed throughout the Middle Ages. With the rise of modern science, however, investigators began to develop explanations of physical phenomena that did not rely on overtly evaluative notions. Many scientists contributed to this development, but chief among them was Galileo. When in 1633 he was forced by the Church to recant, the specific issue was whether the earth goes around the sun. But Galileo's rejection of the geocentric universe was not his most insidious teaching. Far more revolutionary was his use of explanatory categories inconsistent with the Aristotelian tradition.

A good example of this is the explanation of the suction pump first developed by one of Galileo's disciples, Evangelista Toricelli. Practical men had known for a long time that water could be drawn from a well by placing one end of a cylinder in the water and lifting a piston. It was also known that water can be raised only about 32 feet (at sea level) by this method. But *why* does the water rise? And why does it stop rising when it reaches a height of 32 feet?

The Aristotelian explanation depended, first, on the assumption that nature abhors a vacuum. As the piston is raised, a space is created and no air can enter to fill it. This precipitates a little crisis. The only thing available to fill the vacated space is the water; if the water does not rise, a vacuum will result. Because nature abhors vacuums, there is a tendency for the water to rise. But this means that the water must leave its 'proper place', which is down, not up. (The principle of the 'proper place' does the work of gravity; it explains, for example, why objects fall and why water runs downhill.) The two principles therefore contend against one another. Experimentally, we can see that abhorrence-of-vacuums is more powerful, because the water does leave its natural place to prevent the vacuum from forming. However, after the water rises about 32 feet (at sea level), the two forces are in equilibrium. The height to which the water will rise is a measure of the relative power of the two forces.

However quaint this view now appears, it was a serious view held by serious people, and its explanatory power should not be underestimated. It can also explain, for example, why the suction pump is less efficient at higher altitudes. On a mountainside, water can be drawn less than 32 feet, and the higher one goes, the shorter the column of water will be. The Aristotelians knew these facts, and they were not stupid; they would not have accepted an account that could not explain them. But the Aristotelian categories were fully sufficient to explain the phenomena. The explanation was that on a mountainside the water is even farther from its 'proper place', and so the downward pull will be greater.

Nevertheless, Toricelli offered a fundamentally different sort of explanation. He observed that air has weight, and argued that it is the weight of the column of air above the water that accounts for the pump's success. When the piston is raised, the weight of the atmosphere is lifted from the water in the cylinder. But the weight of the air is still pushing down on the water in the well outside the tube; therefore the water flows from the well up into the pump. The water will go no higher than 32 feet because at that point the weight of the water balances the weight of the air; on a mountainside, the column of water is shorter because there is less air above it. The height of the water is a measure, not of the intensity of nature's detestation of vacuums, nor of the strength of the attraction of a 'proper place', but of the weight of the air.

Of course it is Toricelli's explanation that is accepted today; the Aristotelian account is found only in history books. But the change was not simply a matter of substituting one explanation for another; more importantly, it was a matter of adopting a different *kind* of explanation. The Aristotelian account saw nature as governed by principles of value: water leaves its 'proper place' to prevent a vacuum from forming because vacuums are somehow 'abhorrent'. These terms were not mere metaphors or convenient ways of speaking, as comparable terms might be for a modern physicist. As I have already remarked, Aristotle held that everything that happens in nature happens for a purpose; the rains come in order that the plants may grow. In his *Physics* Aristotle carefully considers the alternative view that natural occurrences take place 'not for the sake of something, nor because it is better so', but only as the result of 'necessity'—the blind operation of the law of cause and effect. It would then be a 'coincidence' that the falling rain aids the plants. This possibility Aristotle rejects. His cosmology, like the Christian world-view that would come later, saw nature moving to bring about states of affairs that are 'better' than what would exist if everything were a matter of coincidence. The principles that describe how nature works will, therefore, incorporate conceptions of how things ought to be, as well as notions of how things are. The suction pump operates according to those principles. Toricelli's explanation, on the other hand, accounts for the workings of the pump without recourse to any normative conceptions. That is the deep significance of the new style of explanation.

Biology without Purpose

Toricelli's account of the suction pump is, of course, only one among many examples that could be given. Just two hundred years separated his account from the publication of Darwin's *Origin*; but during that time, Newton had done his great work, and the triumph of the new style of explanation was virtually complete. By the middle of the eighteenth century, the ideal of the physical sciences was to be 'value-free', to understand nature as it is without any assumptions about how it ought to be. In the physical sciences, teleological explanation was out.

Biology, however, was another matter. No way had been found to eliminate the notions of purpose and goal-directedness from the descriptions of biological organisms, and to most scientists, such an idea seemed impossible. Few even tried. Instead, the biological and physical sciences came to be regarded as fundamentally different from one another, requiring different explanatory principles and different styles of investigations. Looking back, we can see that this was an illusion created by the more rapid advancement of the physical sciences. But at the time the differences seemed striking and real. If the Church had lost its battle with Galileo and his ilk, it could still take comfort in the state of biological understanding, which saw organic nature as a grand design of purposeful construction—until, that is, Darwin changed everything.

On Darwin's account of how species come to have their characteristics, conscious design plays no part. When we think of how natural selection operates to preserve characteristics favourable to survival, it is easy to misunderstand what is happening. It should be emphasized that, on Darwin's view, nature does not provide the helpful characteristics in order to benefit the organisms. Protective coloration is not given to the grouse to enable it to avoid the hawk. Nor is the desirable colour produced *in response to* the hawk's threat. The grouses' colours vary, usually in tiny degrees, randomly. They would vary in the same way even if there were no hawks about. The fact that some variations benefit some birds, giving them a better chance to survive, is the merest coincidence.

The same holds for changes that occur in response to changes in the animals' environments. Environmental changes will, according to the theory of natural selection, lead to changes in the characteristics of species. But the environmental change does not cause individual organisms to change; it only means that different characteristics will be selected for. If the environment grows colder, animals with thicker fur will fare better; but no individual animal's fur grows thicker *because* it is colder. The animal's fur will be thicker only because either (*a*) the thickness of its fur has varied randomly, without any regard to environmental conditions, or (*b*) it is the descendant of animals with this lucky variation, who, because of their good luck, were better able to survive the colder conditions and therefore were able to pass on the characteristic to their young. We must not imagine that there is a 'guiding hand' of any sort, except in the most metaphorical sense.

It is useful to think of the variations as being produced merely by chance. (Darwin himself used the term 'chance', but he did not like it; he emphasized that there might be laws unknown to him that govern the production of variations. But in the present context, it is a useful notion, for it stresses the lack of pre-determination between the variations and the purposes some of them happen to serve.) Once they are randomly generated, the variations then form the materials on which natural selection works: if a variation *happens* to confer an advantage, it is preserved; otherwise, it is not. Is it not, then, a miracle that advantageous variations occur? No; the apparent miracle is possible only because many millions of organisms are born over a very long time; the great numbers permit all manner of variations, some of which are bound to help. The outcome seems miraculous only when we take a short view of the matter.

This is the great difference between Darwin's view and the failed theories of the other evolutionists who went before him: Lamarck, for example, still used value-laden notions when he spoke of inner forces directing change towards increased perfection, whereas for Darwin there was only random variation providing the materials on which natural selection can then operate. For Darwin there was nothing in the constitution of any organism that propels its development in any particular direction. Nor were there any 'higher' or 'lower' forms of life; nor any 'progress': there were only organisms adapted in different ways to different environments, by a process ignorant of design or intention.

Darwin and Paley

Darwin considered the theory of natural selection to be a rival of, and a replacement for, the idea that particular aspects of nature were consciously designed. The two notions were, on his view, utterly incompatible. 'The old argument of design in nature,' he said, 'which formerly seemed to me so conclusive, fails, now that the law of natural selection has been discovered.' In the *Origin*, although he tried to side-step the explosive question of human evolution, he did not attempt to avoid the equally critical question of design. In fact, he went out of his way to meet it head-on, when he included in his book a discussion of the evolution of the eye. His readers could not fail to recognize this as the favourite example of William Paley, the foremost theological proponent of the design hypothesis. Paley's book *Evidence of the Existence and Attributes of the Deity* had been read and admired throughout Britain for decades. Everyone knew Paley's argument; Darwin could not have chosen a more widely known or respected opponent.

The human eye is so wonderfully suited to its purpose, in such complex ways, that Paley had cited its very existence as proof that we are the products of divine creation. How else, he argued, can such a marvellous design be explained? Paley claimed that we have exactly the same reason to believe the eye was produced by an intelligent designer as we have to believe that objects such as telescopes are produced by intelligence.

As far as the examination of the instrument goes, there is precisely the same proof that the eye was made for vision, as there is that the telescope was made for assisting it. They are made upon the same principles; both being adjusted to the laws by which the transmission and reflection of rays of light are regulated.

The force of the argument, however, is best felt in the presentation of its details, and Paley was very good at this:

Besides that conformity to optical principles which its internal constitution displays, . . . there is to be seen, in everything belonging to it and about it, an extraordinary degree of care, and anxiety for its preservation, due, if we may so speak, to its value and its tenderness. It is lodged in a strong, deep, bony socket, composed by the juncture of seven different bones, hollowed out at their edges . . . Within this socket it is embedded in fat, of all animal substances the best adapted both to its repose and motion. It is sheltered by the eyebrows; an arch of hair, which like a thatched penthouse, prevents the sweat and moisture of the forehead from running down into it.

But it is still better protected by its lid. Of the superficial parts of the animal frame, I know none which, in its office and structure, is more deserving of attention than the eyelid. It defends the eye; it wipes it; it closes it in sleep. Are there, in any work of art whatever, purposes more evident than those which this organ fulfills? . . .

In order to keep the eye moist and clean (which qualities are necessary to its brightness and its use), a wash is constantly supplied by a secretion for the purpose; and the superfluous brine is conveyed to the nose through a perforation in the bone as large as a goose-quill. When once the fluid has entered the nose, it spreads itself upon the inside of the nostril, and is evaporated by the current of warm air, which, in the course of respiration, is continually passing over it . . . [C]ould the want of the eye generate the gland which produces the tear, or bore the hole by which it is discharged—a hole through a bone?

This is indeed impressive, apt to stir the heart as well as to persuade the mind, and as a young student Darwin had been completely convinced by it. Considered dispassionately, however, the argument Paley advances is full of holes. For one thing, he is simply mistaken when he says that 'there is precisely the same proof' that the eye and the telescope are both products of intelligent creation. We can observe people designing and making telescopes, and so we know for certain how they come to be made. But no one has ever observed a creator designing and making an eye. Therefore there is not 'precisely the same proof' in both cases. This, and much more, was pointed out by David Hume in his *Dialogues Concerning Natural Religion*, published twenty-three years before Paley's book—so Paley has little excuse for having made this error.

Today Hume's book is generally thought to be a definitive refutation of the design argument. (This is one of the few things about which contemporary philosophers agree.) However, like Paley, most readers in the nineteenth century were unimpressed by Hume's logic. More important was the fact that the hypothesis of divine creation provided a way of accounting for the eye, and other apparently purposive elements of nature. Why should people abandon a useful way of understanding when there is none better available? Despite any logical weaknesses that the argument from design might have, the hypothesis could not be robbed of its appeal until an alternative account was supplied. That is what Darwin did, and to underscore the point, he discussed Paley's example.

Darwin also realized that, quite apart from their use as evidence of intelligent design, complex organs such as the eye present a special problem for the theory of natural selection. First, they are problematic because they are constructed of numerous parts, each of which appears to be useless except when working together with the others. How are we to conceive of the evolution of all these parts? Are we to imagine a rudimentary eye, a rudimentary tear-duct, a rudimentary lid, and all the rest, developing alongside one another? And secondly, remember that, on Darwin's view, such complex organs are the result of a great many small variations, that 'add up' to the mature organ after millions of generations of evolutionary modification. Even though it is easy to see that the fully developed eye is useful to its possessor, of what use is a half-eye that still has many generations to go before it will be complete? Why should a half-eye be selected for and preserved for further development?

To deal with these problems, Darwin made two points. First, he emphasized that a bit of anatomy may originally be preserved by natural selection because it serves a different adaptive purpose from the one it eventually comes to serve. Later, it may come to play a part in some complex structure because it just happened to be present. Nature may jury-rig a complex structure out of whatever materials happen to be at hand. Secondly, Darwin called attention to what present-day theorists call *the intensification of function*. In the later stages of its development a biological structure might confer a benefit that in its early stages it conferred to a much lesser degree. To explain the eye, Darwin had to appeal to both these points.

To suppose that the eye, with all its inimitable contrivances for adjusting the focus to different distances, for admitting different amounts of light, and for the correction of spherical and chromatic aberration, could have been formed by natural selection, seems I freely confess, absurd in the highest possible degree. Yet reason tells me, that if numerous gradations from a perfect and complex eye to one very imperfect and simple, each grade being useful to its predecessor, can be shown to exist . . . then the difficulty of believing that a perfect and complex eye could be formed by natural selection, though insuperable by our imagination, can hardly be considered real.

All we have to imagine is that a nerve only slightly sensitive to light confers on an organism some small advantage in the competition for survival; then we can understand the establishment of the first rudimentary eye. From that simple thing will eventually come our complex eyes.

In living bodies, variation will cause the slight alterations, generation will multiply them almost infinitely, and natural selection will pick out with unerring skill each improvement. Let this process go on for millions on millions of years; and during each year on millions of individuals of many kinds; and may we not believe that a living optical instrument might thus be formed as superior to one of glass, as the works of the Creator are to those of man?

And if the eye itself can be formed in this way, then so can the tear-ducts, the eyelid, the bone, and the rest. Take the lid, for example: imagine that a rudimentary eye has been established, and that in some organisms a slight variation has resulted in a small fold of skin that somewhat protects it. The skin is not there *in order to* protect the eye; it was originally developed because it conferred a different benefit. But now that it is there, it can serve this new 'purpose', and this new feature will be selected for, and further modified, in the usual way.

Darwin had answered Paley in the most effective way. Hume and the other philosophical critics of the design argument could point out its logical deficiencies, but they could not supply a *better* way of understanding the apparent design of nature. Taking away design as an explanation, they left nothing in its place; and so it is no wonder that, despite Hume's criticisms, even the brightest people continued to believe in design. Darwin did what Hume could not do: he provided an alternative, giving people something else they could believe. Only then was the design hypothesis really dead.

Theistic Responses, Darwinian Replies

A theist who did not wish to reject evolution might respond to all this in either of two ways. One response would be to argue that, contrary to what Darwin says, evolution by natural selection is not really incompatible with design. A different response would be to say, in effect, 'So what?' Suppose that Darwin *has* shown the design hypothesis to be false: Does it follow that theism is false? No, for it might be argued that the design hypothesis was never a necessary component of theism. Rejecting it, therefore, does not mean that one must reject theism.

Let us consider these theistic responses one at a time.

1. Is Darwinism really incompatible with the idea that the world, and all its inhabitants, are the products of intelligent design? The philosopher George Mavrodes is one of many thinkers who have argued that there is no incompatibility here. Mavrodes distinguishes a 'naturalistic' interpretation of the evolutionary process, according to which the process is 'explicable entirely in terms of natural law without reference to a divine intention or intervention', from a theistic interpretation, according to which 'there was a divine teleology in this process, a divine direction at each crucial stage in accordance with divine plan or intention'. Then he argues that there is no evidence that rules out the theistic interpretation.

If the paleontological evidence really does support the naturalistic version, then we must have some idea of how the bones would have been different if God had been involved in the process. If we don't have that idea, then we have no legitimate way of construing our evidence in support of one of these hypotheses rather than the other. People who take the theistic possibility seriously are not likely, therefore, to be persuaded simply by more bones. But I think they are also unlikely to find any evolutionist who will give them a plausible and well-supported idea of how the evidence would have been different if God were directing the process.

How would the evidence have been different, if God were directing the process? Mavrodes specifically mentions paleontological evidence, but of course the evidence for evolution includes much more than that, so we may take his challenge to include other kinds of evidence as well. It is an important challenge, that goes to the heart of the matter. Mavrodes thinks it 'unlikely' that any evolutionist could meet it. Nevertheless, some have tried. One who tried was Darwin.

Darwin's treatment of this issue was in two parts.

(*a*) First, Darwin argued that the theory of 'descent with modification' does lead to different expectations from the hypothesis of intelligent design. We can, therefore, test the two hypotheses, empirically, by seeing which expectations are realized in fact. For example, we would not expect an intelligent designer to include useless parts in organisms; but, on the hypothesis of evolution by natural selection, we would expect to find such useless parts, because they would be the vestiges of once-useful structures. And things are, in fact, as the evolutionary hypothesis predicts: in humans we find muscles that can no longer move ears, useless body-hair, a vermiform appendix that serves no purpose, the remnants of a tail, and so on. Darwin remarks that 'Those who can persuade themselves that purposeless organs have been specially created, will think little of this fact. Those on the contrary, who believe in the slow modification of organic beings, will feel no surprise that the changes have not always been perfectly effected.'

There is another kind of evidence that Darwin thought counts against design and in favour of natural selection. In his book on orchids, which he wrote immediately after *The Origin of Species*, Darwin emphasizes that anatomical structures that originally served one purpose may later come to serve other ends. 'The regular course of events seems to be', he said, 'that a part which originally served for one purpose, becomes adapted by slow changes for widely different purposes.'

If a man were to make a machine for some special purpose, but were to use old wheels, springs, and pulleys, only slightly altered, the whole machine, with all its parts, might be said to be specially contrived for its present purpose. Thus throughout nature almost every part of each living being has probably served, in a slightly modified condition, for diverse purposes, and has acted in the living machinery of many ancient and distinct specific forms.

The orchid book is full of examples. On one page Darwin observes that nectar, useful in attracting insects, was originally 'an excretion for the sake of getting rid of superfluous matter during the chemical changes which go on in the tissues of plants, especially whilst the sun shines'.

And then he adds: 'It is in perfect accordance with the scheme of nature, as worked out by natural selection, that matter excreted to free the system from superfluous or injurious substances should be utilised for [other] highly useful purposes.'

But couldn't these facts also be 'in perfect accordance with the scheme of nature, as worked out by [the design hypothesis]'? Darwin thought not. His thought was that the two theories would be supported by very different kinds of evidence. Evidence of perfect, elegant adaptation would support the design hypothesis, while evidence of improvised, jury-rigged adaptation would support the evolutionary hypothesis. The contrast is between designs and contrivances; and the latter are what we actually find. Nature rigs its contraptions using whatever materials are at hand, including materials that no rational engineer would ever choose for the new purpose. 'The larvae of certain beetles', Darwin pointedly remarks, 'use *their own excrement* to make an umbrella-like protection for their tender bodies.' The obvious follow-up is to ask whether that is a design we would expect from a perfect engineer with unlimited power and resources.

(*b*) It might be replied, on behalf of the theist, that this misses the point. The Darwinian arguments we have just been considering are directed against the doctrine of special creation, that is, against the idea that God creates each species separately, from scratch. Those arguments do not touch the very different suggestion made by George Mavrodes, namely, that God has worked *through* the evolutionary process: evolution has occurred, just as Darwin says, but there was 'a divine direction at each crucial stage in accordance with the divine plan or intention'. Living beings may be contraptions, but they are God's contraptions. What reasoning justifies dismissing this idea? This brings us to the second part of Darwin's discussion.

As it stands, the mere suggestion of a 'divine direction' is too vague to be tested. If God works through the evolutionary process, how exactly is this supposed to happen? We need a specific proposal. Remember how evolution by natural selection operates. Variations occur, and those that confer advantages in 'the struggle for life' are preserved to be transmitted to future generations. At what point does God intervene in the process, and how? The most obvious conjecture would be that God intervenes by providing the specific variations that he knows will confer advantages. If God wants to modify wolves in a certain direction, for example, he can arrange things so that as the climate grows colder some animals will have slightly thicker fur. If he wants that species to become extinct, he will not provide the thicker fur and let them be killed off by the cold. This is the most natural way to fill in the details of the theistic interpretation, and Darwin addressed it directly in *The Variation of Animals and Plants under Domestication*.

To refute this suggestion Darwin relies on an argument from analogy that harkens back to his discussion of artificial selection. Natural selection, Darwin urged, is closely related to the process by which humans intentionally produce new varieties of plants and animals by selective breeding. Both processes take advantage of naturally occurring variations. If the variations utilized in natural selection are not simply random, but are directed by God, then this is no less true of the variations seized upon by breeders. But Darwin thinks it is impossible to believe that variations occur for the benefit of the breeders, and so he rejects the idea that the variations occur in order that natural selection can take a given direction:

An omniscient Creator must have foreseen every consequence which results from the laws imposed by Him. But can it be reasonably maintained that the Creator intentionally ordered, if we use the words in any ordinary sense, that certain fragments of rock should assume certain shapes so that the builder might erect his edifice? If the various laws which have determined the shape of each fragment were not predetermined for the builder's sake, can it be maintained with any greater probability that He specially ordained for the sake of the breeder each of the innumerable variations in our domestic animals and plants—many of these variations being of no service to man, and not beneficial, far more often injurious, to the creatures themselves? Did He ordain that the crop and tail-feathers of the pigeon would vary in order that the fancier might make his grotesque pouter and fantail breeds? Did he cause the frame and mental qualities of the dog to vary in order that a breed might be formed of indomitable ferocity, with jaws fitted to pin down the bull for man's brutal sport? But if we give up the principle in one case—if we do not admit that the variations of the primeval dog were intentionally guided in order that the greyhound, for instance, that perfect image of symmetry and vigour, might be formed—no shadow of reason can be assigned for the belief that variations, alike in nature and the result of the same general laws, which have been the groundwork through natural selection of the formation of the most perfectly adapted animals in the world, man included, were intentionally and specially guided.

Note that Darwin's conclusion is not that it is impossible that (i) God intentionally provided the variations utilized in natural selection, while (ii) he did not intentionally provide other variations for the sake of the breeders. Darwin's conclusion is only that there is no reason for thinking this, in light of the fact that both kinds of variations are produced in the same way, through the operation of the same natural laws. Here, as always when dealing with matters that touch on religion, Darwin assumes that what is important is determining what it is reasonable to believe. If the theist were to respond that there is nothing *inconsistent* in this combination of beliefs, Darwin could agree, but that would be irrelevant. There are innumerable absurd beliefs that might be held without self-contradiction; that is too weak a test to be of much interest. Of more interest is whether a belief is reasonable.

(*c*) The foregoing discussion was prompted by Mavrodes's challenge to specify the evidence that would rule out a theistic interpretation of the evolutionary process. I have suggested that, given the most natural way of construing the theistic interpretation, Darwin did provide some reason for rejecting it. Perhaps there are other ways of construing that interpretation; if so, other replies would have to be formulated. There is one possibility, though, that deserves special mention.

It is possible to construe the theistic interpretation in such a way as to *guarantee* that there could never be any evidence against it. This could be done by refusing to specify any details at all, leaving it a perfectly open question how God works through the evolutionary process. Given no specific content, the theistic interpretation would imply nothing at all about what nature is like. Then, no matter what discoveries are made, the theist could always say: 'Yes, that is the way things are, and that is the way God planned it.' If one takes this approach, then the challenge to provide evidence against the theistic interpretation can never be met.

I mention this possibility because I think that, when people insist that evolution can be interpreted theistically, some idea like this is more often than not lurking in the background. But ultimately it is an unsatisfactory way of defending the theistic interpretation. Logically it is comparable to the crudest reasoning used by so-called 'scientific creationists' who reject evolution altogether. The most desperate strategy yet devised for denying that evolution has occurred is the suggestion that, when God created the world a few thousand years ago, he could have made the fossils, the geological strata, and so on, at the same time—there never were any dinosaurs, only dinosaur bones. (If various tests seem to show that the bones are really millions of years old, that is only because God created them in such a condition that the tests would give this result.) The creationist can say that this hypothesis is perfectly consistent with all the data, and can smugly challenge the evolutionist to provide evidence that the hypothesis is wrong.

At first glance, it seems that the evolutionist can have no satisfactory reply, because any evidence that might be adduced in support of evolution can be incorporated within this creationist scenario. Nevertheless, it is not difficult to explain what is wrong with the creationist ploy. It is self-defeating in a curious way. Suppose the hypothesis were true. It would mean that, even though the world is only a few thousand years old, God has filled it with evidence that makes it reasonable for us to believe otherwise. (Contrary to Descartes's famous dictum, it turns out that God is indeed a deceiver.) In fact, God has provided evidence that makes it reasonable for us to believe that the earth is 4.6 *billion* years old. And since it is reasonable for us to believe this, it follows that it is reasonable for us to believe that the creationist hypothesis is false.

Much the same can be said about the theistic interpretation of the evolutionary process, when it is construed in the open-ended way that guarantees compatibility with all possible evidence. Suppose God *is* somehow involved in the process that evolutionary biologists since Darwin have been describing. This would mean that he has created a situation in which his own involvement is so totally hidden that the process gives every appearance of operating without any guiding hand at all. In other words, he has created a situation in which it is reasonable for us to believe that he is not involved. But if it is reasonable for us to believe that, then it is reasonable for us to reject the theistic interpretation.

2. But what if Darwin's argument is accepted? What if, on Darwinian grounds, one rejects the design hypothesis? Does it follow that one must also reject theism? That would follow only if the design hypothesis is a necessary component of theism. This brings us to the second main strategy for resisting the sceptical conclusion: it might be said that the design hypothesis, in the form advocated by Paley and criticized by Darwin, was never necessary for theism in the first place.

Can theism be separated from belief in design? It would be a heroic step, because the design hypothesis is not an insignificant component of traditional religious belief. But it can be done, and in fact it has been done, by the eighteenth-century deists. The deists did not believe that the world is looked after by an ever-present, loving God; instead, they viewed God as a creator who made the world, and established the laws by which it would operate, but who thereafter kept hands off. Thus, although God is credited with designing the grand plan of the universe—the laws of nature—he is not seen as concerning himself with the details. On this view, we might say that although God created the mechanisms by which natural selection occurs, he did not design its products in any other sense.

Since deism is a consistent theistic view, it is tempting simply to conclude that theism and Darwinism must be compatible, and to say no more. But the temptation should be resisted, at least until we have made clear what has been given up in the retreat to deism. If religious belief is reduced to this, is it worth having? What remains is a 'God' so abstract, so unconnected with the world, that there is little left in which to believe. In 1927 Freud said about this sort of belief:

In reality these are only attempts at pretending to oneself or to other people that one is still firmly attached to religion, when one has long since cut oneself loose from it. Where questions of religion are concerned, people are guilty of every possible sort of dishonesty and intellectual misdemeanour. Philosophers stretch the meaning of words until they retain scarcely anything of their original sense. They give the name of 'God' to some vague abstraction which they have created for themselves; having done so they can pose before all the world as deists, as believers in God, and they can even boast that they have recognized a higher, purer concept of God, notwithstanding that their God is now nothing more than an insubstantial shadow and no longer the mighty personality of religious doctrines.

To this it might be replied that deism does contain one idea worth preserving, even if it banishes the 'mighty personality of religious doctrines', namely this: it retains the idea of God as the ultimate cause of the universe. But Darwin himself made the right observation here. Since no one knows what the ultimate cause of the universe was, to say that it was God is the merest speculation. One can *say* that, but one can give no good reason in its support. Darwin thought that the more honest approach is to admit one's ignorance and remain silent. Surely he was right.

Moreover, deism represents the retreat of religious belief in still another sense. There is now far less *content* to the idea of God. The concept of God as a loving, all-powerful person, who created us, who has a plan for us, who issues commandments, and who is ready to receive us into Heaven, is a substantial concept, rich in meaning and significance for human life. But if we take away all this, and leave only the idea of an original cause, it is questionable whether the same word should even be used. By keeping the original word, we delude ourselves into thinking that we are talking about the same thing. We may even, as Freud says, 'boast that [we] have recognized a higher, purer concept of God'—but the boast might well deserve the scorn that Freud heaps upon it.

In summary, then, the atheistical conclusion can be resisted, but only at great cost. The concept of God that survives is so vague that it has little use in explaining either nature in general or human nature in particular. God has retreated so far from the world we know that he has become, in Freud's words, 'nothing more than an insubstantial shadow'.

Conclusion

Must a Darwinian be sceptical about religion? If by this we mean 'Is it logically possible to be both a Darwinian and a theist?' then the answer is that it is possible. There is nothing in Darwinism that proves that every form of theism must be false. But the question might have a somewhat broader meaning. Even if Darwinism does not prove theism is false, it still might provide powerful reasons for doubting its truth.

An evolutionary perspective undermines religious belief by removing some of the grounds that previously supported it. Gould says that science 'doesn't intersect the concerns of theology'. Surely that is wrong; science and theology may have different concerns, but they do intersect. The most important point of intersection has to do with purposive explanations of natural phenomena. For theology it is no small matter whether nature is interpreted teleologically. When the world is interpreted non-teleologically—when God is no longer necessary to explain things—then theology is diminished.

It is often said that the theologian's explanations are different in kind from the scientist's explanations, and that the two can co-exist alongside one another. It is not so often noted that mere 'co-existence' is a great come-down for theology. The history of religion in the age of science is the story of religion's steady retreat from a central place in our way of understanding the world to a place that is vanishingly small. Traditional religion drew strength from the fact that theological explanations were needed to make sense of the world. So long as people had no other way of explaining why the world is as it is, the grip of religious conceptions was powerful: everyone, scientists included, had reason to believe. But after Darwin, with teleological conceptions banished from our understanding of nature, there is markedly less work for religion to do, and God looks more and more like an unnecessary hypothesis. That is why Darwinism, even if it does not make theism impossible, still makes it far less attractive than ever before.

Finally, we may consider the implications of all this for the overall argument of this book. Our primary interest here is in the support that religion has traditionally provided for the doctrine of human dignity. The lives and interests of human beings are morally important, it is said, in a way that the lives and interests of other animals are not, because humans are made in the image of God and have a special place in the divine order. I have argued in this chapter that Darwinism undermines theism. Some readers might go part way with the argument, but stop short of concluding that all forms of theism must be rejected. They might conclude instead that some suitably refined version of theism is still tenable. The question that will remain, however, is whether that refined theism is sufficiently robust to support the image of God thesis. The image of God thesis does not go along with just any theistic view. It requires a theism that sees God as actively designing man and the world as a home for man. If, by abolishing the view of nature as designed in substantial detail, Darwinism forces a retreat to something like deism, then we are deprived of the idea that man has a special place in the divine order. Even if we can still view nature as in some sense God's creation, we will no longer have a theism that supports the doctrine of human dignity.

4 How Different are Humans from Other Animals?

idea of the Rationality thesis

IT has always been difficult for humans to think objectively about the nature of non-human animals. One problem, frequently remarked upon, is that we tend to anthropomorphize nature and see animals as too much like ourselves. An opposite but less frequently noticed difficulty is connected with the fact that, even as we try to think objectively about what animals are like, we are burdened with the need to justify our moral relations with them. We kill animals for food; we use them as experimental subjects in laboratories; we exploit them as sources of raw materials such as leather and wool; we keep them as work animals—the list goes on and on. These practices are to our advantage, and we intend to continue them. Thus, when we think about what the animals are like, we are motivated to conceive of them in ways that are compatible with treating them in these ways. If animals are conceived as intelligent, sensitive beings, these ways of treating them might seem monstrous. So humans have reason to resist thinking of them as intelligent or sensitive.

At times this resistance has taken an extreme form. In the seventeenth century, a time of rapid and exciting advancement for physiology, the functioning of the major organs and the circulation of the blood were beginning to be understood for the first time. But these advances were achieved through experimental procedures that subjected animals to excruciating tortures. Dogs, for example, would be restrained by nailing their paws to boards, and then would be cut open so that the working of their innards could be observed. This was long before the development of anaesthetics, and the dogs' vocal chords would sometimes be cut so that their shrieks would not disturb the anatomists.

If any scientist was troubled, traditional moral doctrines offered reassurance. Thinkers such as Aquinas had long taught that 'Charity does not extend to irrational creatures', and that brute animals exist only for the use of men. Such reassurances might, however, have seemed insufficient: after all, regardless of whether the animals were rational, and regardless of whether they were intended for human use, the fact remained that they were being made to suffer terribly. That alone would be enough to give one pause. To remove *all* reason for guilt, it would be necessary to view the animals as incapable of feeling pain. And, strange as it might seem today, that is just how they came to be viewed. Mere animals, it was said, are so different from humans that they cannot even feel pain.

René Descartes (1596–1650), commonly called 'the father of modern philosophy', was the chief proponent of this view. Descartes held that the mind and the body are radically different types of entities. The mind is wholly immaterial in nature, whereas the body is simply a kind of machine. Humans, because they have minds as well as bodies, are capable of thought and feeling. But mere animals, Descartes said, have no minds. They are, therefore, nothing but machines, incapable of any sort of conscious state, including pain. 'My opinion', he wrote, 'is not so much cruel to animals as indulgent to men . . . since it absolves them from the suspicion of crime when they eat or kill animals.' It was also indulgent to physiologists who no longer needed to be concerned about the 'sufferings' of their animal subjects. In at least some laboratories, Descartes's view was adopted enthusiastically. Nicholas Fontaine wrote in his memoirs, published in 1738, about a visit to one such laboratory:

They administered beatings to dogs with perfect indifference, and made fun of those who pitied the creatures as if they felt pain. They said the animals were clocks; that the cries they emitted when struck were only the noise of a little spring that had been touched, but that the whole body was without feeling. They nailed poor animals up on boards by their four paws to vivisect them and see the circulation of the blood which was a great subject of conversation.

There was another reason why pre-Darwinian thinkers found it convenient to deny that animals have any significant psychological capacities: the suffering of animals presented a serious theological problem. A just and all-powerful God would not create beings to suffer for no purpose. The suffering of humans could be explained (or so it was thought) by associating it with the doctrine of the Fall of Man: human suffering is the consequence of Adam's sin. But animals are not descended from Adam, and have no share in Original Sin; moreover, they have no hope of Heaven, which could redeem their earthly suffering. Thus the pain of animals is apparently an intractable theological problem. Darwin, as we have seen, thought about this problem and concluded that God probably does not exist. But what solution was there for those unwilling (or unable) to draw this conclusion? The solution favoured by many was to deny, against all apparent indications, that animals suffer. Malebranche, a contemporary of Descartes, welcomed

Descartes's view for this reason. This theological idea was still current when Darwin published *The Origin of Species*, and some readers objected to Darwin's theory on the grounds that if he were right it would mean that animals must suffer.

It is easy today, looking back, to think Descartes's view ridiculous. How could anyone seriously believe that animals do not feel pain? After all, we have virtually the same evidence for animal pain that we have for human pain. When humans are tortured, they cry out; so do animals. When humans are faced with painful stimuli, they draw back and try to escape; so do animals. Pain in humans is associated with the operation of a complex nervous system; so it is with animals. The only indication of pain in humans that we do not have for animals is that humans can tell us, in words, that they are suffering. But this is not true of *all* humans; infants cannot speak, and neither can some severely retarded or senile people—yet we do not doubt that they suffer when they are hurt. So, on what grounds could anyone possibly say that animals are insensitive to pain?

Descartes's view was extreme, even for his own time, and despite its wide influence most thinkers did not share it. Nevertheless, it was a possible view then, in a way that it is not possible now. The reason Descartes's view of animals is not possible today—the reason his view seems so *obviously* wrong to us—is that between him and us came Darwin. Once we see the other animals as our kin, we have little choice but to see their condition as analogous to our own. Darwin stressed that, in an important sense, their nervous systems, their behaviours, their cries, *are* our nervous systems, our behaviours, and our cries, with only a little modification. They are our common property because we inherited them from the same ancestors. Not knowing this, Descartes was free to postulate far greater differences between humans and non-humans than is possible for us.

It would be a mistake, however, to think that people now generally accept, without qualification, that animals can suffer pain. Even today this small concession comes hard; vestiges of Cartesianism remain. One indication is the common practice, in both scientific writing and in the popular press, of placing words like 'pain' and 'suffering' in quotation marks when they are attributed to animals, as though the concepts have only doubtful application. In 1986 the *Birmingham News* ran a story about a government study of experimentation under this headline: 'Researchers said to be minimizing "suffering" of laboratory animals.' The newspaper's readers might have been happy to learn this, but they were also reassured, if only subliminally, by the scare-quotes.

Another common ploy is, while not denying that it is possible in theory for animals to suffer, to be sceptical about whether they do so in any particular case. In 1978 the Government of India stopped exporting rhesus monkeys to the United States because the US had violated a provision of the export agreement which forbade using the animals in nuclear weapons research. The Defense Nuclear Agency confirmed that, over a five-year period, 1,379 primates—nearly all of them rhesus monkeys—had been used in its tests. In one set of tests, the animals had been subjected to lethal doses of radiation and then forced by electric shock to run on a treadmill until they collapsed. Before dying, the unanaesthetized monkeys suffered the predictable effects of excessive radiation, including vomiting and diarrhoea. After acknowledging all this, a DNA spokesman commented: 'To the best of our knowledge, the animals experience no pain.'

Darwin's theory implies that non-human animals—at least, vast numbers of them—not only suffer pain, but are in many other respects intelligent and sensitive beings. He knew that people would be reluctant to accept this, and he had his own explanation of why: exactly 100 years after Fontaine's memoirs were published, Darwin wrote: 'Animals whom we have made our slaves, we do not like to consider our equals.'

ARE HUMANS THE ONLY RATIONAL ANIMALS?

We have already observed that some of Darwin's contemporaries, such as Wallace, believed that although natural selection might explain much about human evolution, it could not account for the development of man's higher intellectual capacities. They therefore argued that evolutionary theory must be supplemented by a limited doctrine of special creation—God must have intervened in history, at some point, to endow humans with rational souls. Darwin, of course, held that the emergence of rational capacities could be explained by the same principles that explain everything else. Human abilities require no special treatment.

Darwin's arguments concerning the rationality of animals must be viewed in this context. He wanted to show that humans are entirely the products of natural selection; but to prove this it was not enough to cite evidence concerning the evolution of man's physical characteristics. Man, it was said, is first and foremost 'the rational animal'; thus a demonstration of man's origins must also include evidence that his rational capacities are the products of natural selection. Part of Darwin's argument was that we find similar rational capacities in other animals; echoing the language of the Cartesians, he rejected the idea that animals are merely 'animated machines'. But if the existence of these capacities in other animals is explained by natural selection, then why not say the same for the human capacities? Darwin did not deny that human rational abilities far exceed those of other animals. But he insisted that the difference is only one of degree, not of kind. 'There is no fundamental difference', he said, 'between man and the higher mammals in their mental faculties.'

From Monkeys to Worms

Darwin is not content simply to argue that animals reason. With characteristic thoroughness, he also suggests that they experience (to greater or lesser degrees) anxiety, grief, dejection, despair, joy, love, 'tender feelings', devotion, ill-temper, sulkiness, determination, hatred, anger, disdain, contempt, disgust, guilt, pride, helplessness, patience, surprise, astonishment, fear, horror, shame, shyness, and modesty—and he supports each suggestion with detailed analysis. Nonetheless, he acknowledges that the central question is about reason. He writes:

Of all the faculties of the human mind, it will, I presume, be admitted that *Reason* stands at the summit. Few persons any longer dispute that animals possess some power of reasoning. Animals may constantly be seen to pause, deliberate, and resolve. It is a significant fact, that the more the habits of any particular animal are studied by the naturalist, the more he attributes to reason and the less to unlearnt instincts.

Darwin gives a number of examples to secure the point. In fact, his technique seems to be to overwhelm the reader with examples, as if to say: if you don't like this one, here's another that you might like better. Some of the examples are from his own observations, but many are borrowed from the writings of other naturalists. The following are typical:

So many facts have been recorded in various works shewing that animals possess some degree of reason, that I will here give only two or three instances, authenticated by Rengger, and relating to American monkeys, which stand low in their order. He states that when he first gave eggs to his monkeys, they smashed them and thus lost much of their contents; afterwards they gently hit one end against some hard body, and picked off the bits of shell with their fingers. After cutting themselves only once with any sharp tool, they would not touch it again, or would handle it with the greatest care. Lumps of sugar were often given to them wrapped up in paper; and Rengger sometimes put a live wasp in the paper, so that in hastily unfolding it they got stung; after this had once happened, they always first held the packet to their ears to detect any movement within. Anyone who is not convinced by such facts as these, and by what he may observe with his own dogs, that animals can reason, would not be convinced by anything I could add.

Despite this last remark, he does add many more examples.

It is easiest to see the behaviour of primates such as monkeys as rational, because they are closely related to us. Darwin, however, was willing to extend the accolade to a great variety of other animals, including some that stretch credibility to the breaking-point. 'Some animals extremely low in the scale apparently display a certain amount of reason', he wrote in *The Descent of Man*; and in his final book *The Formation of Vegetable Mould, through the Action of Worms*, he argued at length that even the lowly earthworm takes some actions as a result of reason, not instinct—'a result', he says, 'which has surprised me more than anything else in regard to worms'.

As I was led to keep in my study many months worms in pots filled with earth, I became interested in them, and wished to learn how far they acted consciously, and how much mental power they displayed. I was the more desirous to learn something on this head, as few observations of this kind have been made, as far as I know, of animals so low in the scale of organization and so poorly provided with sense-organs, as are earth-worms.

It is not surprising that others had not studied the matter—it is difficult to imagine anyone other than Darwin even conceiving of such a project. 'The mental powers of worms'? It sounds like a joke.

The announced purpose of the worm book was to provide evidence that worms are responsible for the formation of the upper layer of the soil, known as the 'vegetable mould'. Darwin had first argued this, as a young man of 28, in a paper read to the Geological Society. It was a significant discovery, revealing something important about the composition of the earth: soil is passed through the intestines of worms, where it is ground up and nutrients are extracted, and then this processed soil is 'cast' to the surface. Darwin proved that from one to two inches of topsoil are produced in this way each decade. But the book had two sub-themes that were of even greater importance. First, the process was an impressive illustration of the way in which large-scale effects may be produced as the accumulation of millions of tiny actions over a very long period of time. By demonstrating this, Darwin added a little more plausibility to one to the great underlying themes of evolution. And secondly, in arguing that the worms are guided by 'a certain amount of reason', Darwin was highlighting, in an especially provocative way, his message about our persistent misunderstanding of the non-human world.

'The mental powers of worms' may sound like a joke, but Darwin was perfectly serious. How, exactly, did the worms display rationality? In observing their habits, Darwin had noticed that they would seize leaves and drag them to their holes, 'not only to serve as food, but for plugging up the mouths of their burrows'. It struck him as remarkable that the worms would usually do this in an efficient way, gripping the leaves by their pointed tips, although he found that in a small minority of cases the worms would try the less effective methods of gripping the leaves by their stalks or in the middle. Is this, Darwin wondered, the result of intelligence? He devotes 35 pages to the question.

Darwin's method was to consider whether hypotheses other than intelligence might explain the worms' behaviour. If other hypotheses could be eliminated, leaving only intelligence as an explanation, then he would have to accept that explanation no matter how strange it seemed. One hypothesis might be that the worms proceed merely by trial and error, learning nothing from their experience. Another might be that they act purely by instinct. To test these hypotheses, Darwin set problems for the worms: he took away the familiar leaves and forced them to try and cope with oddly-shaped leaves that were not native to their region, and with bits of paper cut into different shapes. If the worms were going on nothing but fixed instinct, they should not be able to cope with these new materials at all. But they did manage, quite handily. Moreover, they did so in a sensible way that could not be accounted for as dumb trial-and-error. The worms seemed to be reacting intelligently to their perceptions of the shapes of the new objects. As a result of these observations, Darwin became convinced that the hypotheses of 'unlearnt instinct' and trial-and-error are indefensible, and that 'One alternative alone is left, namely, that worms, although standing low in the scale of organization, possess some degree of intelligence.'

It should be noted, however, that Darwin's brief in behalf of worms was not part of some general campaign to attribute intelligence to *all* creatures, no matter how lowly. He was far too cautious for that. He regarded the matter as an open question, to be decided experimentally in each case. Darwin observed that other lowly animals do not show the same degree of intelligence as the worm. He also examined the habits of the sphex wasp, an insect whose behavioural repertoire includes some types of action similar to the worm's. The sphex drags paralysed grasshoppers into its burrow by gripping the grasshopper's antennae. But the sphex, unlike the worm, is unable to cope with altered situations. When the grasshoppers' antennae are cut off, the sphex is baffled and cannot find alternative ways to accomplish the mission. 'The sphex had not intelligence enough to seize one of the six legs or the ovipositor of the grasshopper, which ... would have served equally well.' The sphex could not solve problems by adapting its behaviour to meet new challenges, and so Darwin concluded that its behaviour, unlike that of the earthworm, was not intelligent.

The worm book, published the year before Darwin's death, was his last major work. Many of its first readers found the attribution of intelligence to worms incredible, and a century later readers might have same reaction. Surely, one might think, he was on much firmer ground with the monkeys and dogs. To take seriously 'the mental powers of worms' suggests that Darwin had become an elderly eccentric. He conceded that his conclusions would 'strike every one as very improbable', and that was putting it mildly. Yet I have emphasized his discussion of worms because it illustrates, even more vividly than his other discussions, some pervasive Darwinian themes.

First, intelligence is not, for Darwin, an all-or-nothing thing that one either has fully or lacks completely. As with other characteristics, we may expect that it will be found distributed in varying degrees all across the animal kingdom. Man is not *the* rational animal; he is merely more rational that the other animals, who have 'a certain amount of reason' also. What better way to demonstrate the continuities in nature, from the lowest creatures to the highest, than by finding a modicum of intelligence even in the earthworm?

Secondly, the presence of intelligence is an empirical matter, to be determined by observation and experiment rather than by reliance upon preconceived notions. Darwin was keenly aware that, although most people have a great many beliefs about what animals are like—beliefs fostered more by philosophy and religion than by impartial observation —they have little real knowledge; and he also knew that, when we look at the facts with an open mind, there are constant surprises. What better way to illustrate the surprises than by considering the worm, a creature we are sure can have no intelligence at all?

Finally, there is the implicit warning against human chauvinism that runs through all Darwin's writing about animal intelligence: we should not apply to other animals a higher standard than we would apply to ourselves.

If worms have the power of acquiring some notion, however rude, of the shape of an object and of their burrows, as seems to be the case, they deserve to be called intelligent; for they then act in nearly the same manner as would a man under similar circumstances.

Where we find animal behaviour that is closely analogous to what we would expect from humans in similar circumstances, and where there are no experimental grounds for distinguishing between them, the animals must be regarded as intelligent, to at least some degree, if humans are so regarded. Anything else, Darwin thought, is illogical and unfair. The best proof we have of the seriousness he attached to this principle is that he would not depart from it even in the case of worms. Considered in this light, Darwin's discussion of 'the mental powers of worms' turns out to be not just the crankish musing of an old man, but a telling choice of example.

Rationality, Language, and Intelligent Behaviour

Looking back on the nineteenth-century naturalists, some commentators have suggested that the Age of Darwin was also the Age of Naïve Anthropomorphism. If Descartes and his followers conceded too little cognitive capacity to the animals, they say, Darwin and his followers erred by going too far in the opposite direction. The psychologist Georgina Ferry remarks that 'These Victorian ladies and gentlemen . . . festooned with collecting tins and encumbered with anthropomorphic preconceptions' were far too willing to 'interpret the observations they collected so assiduously with reference to their own experience'. Wishing to avoid naïve anthropomorphism, many twentieth-century investigators have taken care to emphasize the *differences* between humans and other animals. Thus they have offered a number of arguments to distinguish human rationality from the mere 'pseudo-rationality' of animals. The result is an approach that would please St George Jackson Mivart more than Darwin: the other animals might have bodies similar to ours, but our minds remain something special.

One familiar argument has to do with human linguistic ability. Because we are masters of a complex language, we can formulate thoughts, draw inferences, and in general understand ourselves and what is going on around us in a sophisticated way. Animals who lack a language, the argument says, necessarily lack these associated abilities, and so they cannot be said to be 'rational' in the same sense.

Darwin, however, denies both the assumptions of this argument: he denies that our language is radically different from what we find in non-humans, and he denies that having a language is necessary for being rational.

First, he argues that our use of language differs in degree, not in kind, from the systems of signals used by other animals. Our language, Darwin thought, is probably just the natural extension of some such primitive system:

I cannot doubt that language owes its origin to the imitation and modification of various natural sounds, the voices of other animals, and man's own instinctive cries, aided by signs and gestures . . . we may conclude from a widely-spread analogy, that this power would have been especially exerted during the courtship of the sexes,—would have expressed various emotions, such as love, jealousy, triumph,—and would have served as a challenge to rivals. It is, therefore, probable that the imitation of musical cries by articulate sounds may have given rise to words expressive of various complex emotions . . . [M]ay not some unusually wise ape-like animal have imitated the growl of a beast of prey, and thus told his fellow-monkeys the nature of the expected danger? This would have been a first step in the formation of a language.

But these remarks stop short of the really interesting issues. Darwin assumes that, if the origins of human language can be explained thusly, there is no problem in understanding how further evolutionary development can augment the primitive signal-system until we eventually come to modern English or Hungarian or Chinese. Huxley had pointed out, however, that human language gives men and women capacities that no other animal can even approximate. It enables each generation of humans to pass on to the next their knowledge and experience, so that we have available to us the accumulated wisdom of our forebears. To this it might be added that human language provides for our use a rich system of concepts that non-humans cannot even begin to grasp. Considering this, Darwin's assertion that the difference is only one of degree, not of kind, seems feeble.

How extensive, then, are the linguistic capacities of non-humans? A lot depends on what one means by 'language'. If we mean only the ability to communicate by using conventional signs, animals certainly do that, frequently and competently. ('Any one who has watched monkeys', Darwin says, 'will not doubt that they perfectly understand each other's gestures and expression.') But one might mean something more. Human language has syntactical rules that permit the formation of an indefinite number of new sentences, expressing new thoughts, that have never appeared before. And despite the fact that they have never encountered these sentences before, humans are able to understand them instantly. In a famous phrase of von Humboldt, human language 'makes infinite use of finite means'.

Animal communication involves nothing comparable to the syntactical structures of human language, and the classic view, expressed by Descartes, was that this is what distinguishes man from the beasts:

By [this method] we may also recognize the difference that exists between men and brutes. For it is a very remarkable fact that there are none so depraved and stupid, without even excepting idiots, that they cannot arrange different words together, forming of them a statement by which they make known their thoughts; while, on the other hand, there is no other animal, however perfect and fortunately circumstanced it may be, which can do the same.

And according to Descartes, this in turn means that humans and other animals must have radically different *natures*:

It is not credible that a monkey or a parrot, selected as the most perfect of its species, should not in these matters equal the stupidest child to be found, unless in the case of the brute the soul were of an entirely different nature from ours.

There have been sporadic attempts to show that Descartes was wrong, and that non-humans are capable of mastering a syntactically complicated language, but the results have been disappointing. In the 1960s and 1970s there was excitement about a chimpanzee called Washoe who was taught to use American Sign Language, the sign-system used by deaf people. Washoe, it was said, could ask and answer questions and even improvise phrases that she had not previously been taught. Washoe was trained at the University of Nevada by Roger Fouts; following Fouts's well-publicized 'success', other researchers began working with other primates along similar lines. In California David Premack used coloured tiles to represent words, and reported great success with a chimp named Sarah; while in Georgia Duane Rumbaugh taught another chimp, Lana, to 'speak' by manipulating a computer console. Most astounding of all were claims made on behalf of a gorilla named Koko (trained by psychologist Penny Patterson at Stanford), who was said to joke, invent metaphors, and compose verses.

For a while it seemed that we were on the verge of something truly revolutionary: genuine conversations with members of other species. But these hopes have not, thus far, been realized. After more than twenty years of work, there is still no animal that can converse with humans in any meaningful sense. Moreover, impressive evidence has been presented that the early 'successes' were nothing more than animals responding, uncomprehendingly, to cues unwittingly provided by their human trainers, who would then misrepresent the animals' performances by isolating their few 'meaningful' responses while ignoring the far more numerous instances of gibberish. It had long been known that pigeons can be taught to secure food by pecking coloured buttons in the right order. To substitute inscriptions for colours, so that the pigeon pecks please-give-me-food (rather than red-blue-green-yellow) would create the illusion but not the reality of language. The experiments with the chimpanzees, the critics argued persuasively, did little more.

It seems then, that barring more impressive evidence than is presently available, we are stuck with the conclusion that the linguistic capacities of non-humans are far inferior to those of men and women. Non-humans can communicate with one another, but without anything like a syntactically complex language. This result is disappointing for

Darwin's project of finding continuities, rather than sharp breaks, across species, for this does seem to be just the sort of sharp break that might be appealed to in order to set man apart from other animals. Yet, despite what was said earlier, Darwin was well aware of the vast difference between human language and the communicative devices available to non-humans, and he offered a conjecture about what causes this difference: the development of language, he argued, must depend on the development of the brain; and among the species that survive today, only man's brain is developed in the right way.

As the voice was used more and more, the vocal organs would have been strengthened and perfected . . . But the relation between the continued use of language and the development of the brain has no doubt been far more important. The mental powers in some early progenitor of man must have been more highly developed than in any existing ape, before even the most imperfect form of speech could have come into use; but we may confidently believe that the continued use and advancement of this power would have reacted on the mind by enabling and encouraging it to carry on long trains of thoughts.

Does this mean that humans, but not other animals, are rational?
   'Rationality', like 'language', is a concept that can be interpreted in various ways, and it makes a difference what one takes rationality to be. In one sense, to be rational is to be capable of constructing and following complex chains of reasoning. In this sense Sherlock Holmes was a model of rationality, as are mathematicians. This kind of rationality does seem to depend on possession of a language—as Darwin remarked, 'A long and complex train of thought can no more be carried on without the aid of words, whether spoken or silent, than a long calculation without the use of figures or algebra'—and it is clear enough that humans can 'reason' in this way while other animals cannot.
   On the other hand, rationality is also shown when animals—human or non-human—are able to adjust their behaviour to the demands of the environment in a complex, intelligent way. More specifically, we act rationally when we make choices that are appropriately motivated by our beliefs and attitudes. If we want X, and realize that by doing Y we can get X, and act accordingly, then our behaviour is rational. We act 'for a reason'. If, in addition, we are able to *improvise*, by responding to previously unexperienced environmental conditions, manipulating them to get what we want, the case for attributing rationality is strengthened.
   Darwin seems to have had this latter sense of 'rational' in mind when he expressed scepticism about the importance of language. Are animals rational? In his early notebooks, Darwin made a note to himself: 'Forget

the use of language and judge only by what you see.' When we look at the *behaviour* of non-human animals, he thought, it often shows reason, regardless of whether the use of language is involved. He gives this example:

The orang in the Eastern islands, and the chimpanzee in Africa, build platforms on which they sleep; and as both species follow the same habit, it might be argued that this was due to instinct, but we cannot feel sure that it is not the result of both animals having similar wants and possessing similar powers of reasoning.

Or, as we might say: our best theory of animal behaviour involves attributing to them desires and beliefs. Desires and beliefs, taken together, form reasons for action. Thus, when we explain the animal's behaviour in this way—the animal wants certain things, and realizes that by taking certain steps it can get what it wants—we are seeing its conduct as rational. Sceptics, however, have objected to this easy attribution of rationality. Two types of objection have been offered.

1. Some psychologists have contended that our best theory of animal behaviour does *not* involve attributing such things as desires and beliefs to them. On the contrary, they say, this is a folk-mythical notion that close study exposes as naïve. B. F. Skinner, the renowned behaviourist, has done as much as anyone to cast doubt on such a picture of animal behaviour. Skinner objects strenuously to the use of 'mentalistic' notions in explaining why animals behave as they do. The use of such terms is natural but erroneous. In one essay he described how he would expose this error for students: 'In a demonstration experiment a hungry pigeon was conditioned to turn around in a clockwise direction. A final, smoothly executed pattern of behaviour was shaped by reinforcing successive approximations with food.' Students were then invited to describe what was happening, and they invariably responded with such statements as 'The pigeon *observed* that a certain behaviour seemed to produce a certain result', or 'The pigeon *felt* that food would be given it because of its action'. The students, in other words, explained the pigeon's behaviour as a product of its beliefs and desires. But then, when the origin of the behaviour was revealed, this explanation was exposed as merely fanciful: in reality, the pigeon was only reacting mechanistically to prior conditioning. Skinner believes that, once we realize that all behaviour is similarly caused, we will be forced to conclude that mentalistic explanations are not appropriate.

But this is not just a thesis about *non-human* behaviour. If animal behaviour is shaped by this type of conditioning, so is the behaviour of humans. The conditioning that produces human behaviour is so complex that it is impossible to map, and so we may have the illusion that human behaviour is different. But it is not. Mentalistic explanations are equally inappropriate for human behaviour. The pigeon is only the human writ small.

Two questions naturally arise. First, is Skinner's view correct? And secondly, if it is correct, does it spell trouble for Darwin's view? Of course the overall assessment of Skinnerian behaviourism is too vast a project to be undertaken here. But for present purposes it is enough to mention one central point. Skinner's argument assumes that mechanistic explanations and mentalistic explanations are incompatible—he assumes that if a bit of behaviour can be explained as the product of conditioning, then it will be inappropriate to explain that same behaviour as prompted by desires or beliefs. But that is by no means obvious. Suppose that, as a child, you were rewarded for eating strawberry ice cream and punished for eating vanilla. (Perhaps your parents were psychologists who did this deliberately as an experiment.) As a result, you developed a strong aversion to vanilla. Now, as an adult, when you are offered ice cream, with various flavours available, you invariably refuse the vanilla and take the strawberry. Wouldn't it be true *both* that your action was the product of conditioning (a mechanistic explanation), *and* that your action was prompted by your desires (a mentalistic explanation)? After all, the fact that you now prefer strawberry can itself be explained by the fact that you were conditioned to have this preference. So there is no apparent incompatibility between the two.

A Skinnerian might reply that this argument defends one questionable notion by appealing to an even more questionable one. We are invited to imagine that the causal chain runs from the conditioning to the desire to the action. But this middle term—the desire—is a theoretical idler that does no work. When we look at behaviour and its antecedents, we observe the process of conditioning and we observe the behaviour; and the simple, straightforward hypothesis is that the one causes the other. There is no need to posit some ghostly, unobservable third thing—the 'desire'—as a bridge between them. Prescientific 'common sense' might sanction the use of such notions, but science is more rigorous. In explaining behaviour scientifically, reference to mentalistic entities are not needed and therefore they should be eliminated.

This reply raises large theoretical issues, having to do with the nature of theoretical terms, that go far beyond anything that could be proven by reference to pigeons—or, for that matter, by reference to all the examples of behaviour in the world. To settle those issues we would have to produce a complete philosophy of science. Here we are interested only in a smaller point. The only point we need to make is that mechanistic and mentalistic explanations are not *incompatible*—one does not, in and of itself, rule out the other. A psychologist might eschew mentalistic explanations if he or she has general philosophical views about their eliminability. Skinner certainly has such philosophical views. But that is a different matter: it is one thing to say, on general grounds, the mentalistic terms have no place in a scientific psychology; it is quite another thing to say that mentalistic and mechanistic explanations cannot coexist because of something in their very natures. The former may or may not be correct; the latter is surely wrong.

In the case of Skinner's pigeons, it may seem that once we have the mechanistic explanation, mentalistic explanations are no longer appropriate, but it may seem so only because the behaviour being considered is so simple and inflexible. If Skinner's students had been given a different example, their response might well have been different. Suppose they had been shown someone choosing strawberry over vanilla ice cream and were asked to explain this. Their explanation could have been 'He likes strawberry more than vanilla'. Then, when the history of conditioning was revealed, they might have found that history interesting, but it is doubtful that they would have thought it necessary, just on those grounds, to retract what they had already said. It is the flexibility of human behaviour (a flexibility shared by the behaviour of many non-humans) that seems to make the difference. I shall have more to say about the significance of flexibility below.

At any rate, the question remains whether any of this matters for Darwin's project of seeing human mental powers as similar to the mental powers of animals. In one way, it matters a great deal. Darwin believed that, like humans, other animals have beliefs and desires. If Skinner is right, animals do not have beliefs and desires, and so Darwin was wrong. But in another sense, it doesn't matter at all, because if Skinner is right, then humans don't have beliefs and desires either. Darwinism at the deepest level only insists that human psychology and animal psychology are continuous; what is said about one must be said about the other, excepting only differences of degree. By holding that human behaviour and animal behaviour are *both* merely the products of conditioning, Skinnerian psychology leaves this deeper Darwinian thesis untouched.

2. Another, closely related strategy for undermining confidence in animal rationality has to do with the study of *tropisms*. A tropism is defined by Dean Wooldridge as 'an automatic response differing from other reflexes only in that it affects the movement of the complete organism'. The knee-jerk reflex is a familiar example of automatic, unthinking movement: a stimulus (a blow to the knee) triggers a response (the lower leg moves), without any sort of thought or intention

mediating the two. The idea is that so-called rational behaviour in lower animals is really like that. We may mistakenly believe that an animal is behaving rationally, but in reality the animal's movement is only a machinelike, unreasoning response to a stimulus.

Examples are easy to come by. A bee, after finding food, will return to its hive and perform a 'dance' which informs the other bees of the food's quantity, direction, and distance. Humans have often marvelled at the 'rationality' of this bit of apian behaviour. Yet, when her antennae are stimulated properly, the worker will execute her dance in just the same way, *even though there are no other bees present.* Thus the behaviour is exposed as tropistic; it is merely a mechanical performance in response to a stimulus, and not 'rational' at all.

Or to take a different kind of example, when the male nocturnal moth mates with the female, we may be tempted to see in his evident zest something analogous to the lusty desire of a man—yet it turns out that the male moth's behaviour is triggered entirely by the odour produced by two scent organs near the female's abdomen. When these are removed the male will attempt to mate with the organs, ignoring the nearby female. Again, upon analysis, we seem to find nothing but a fixed, unreasoning response to a stimulus.

Darwin, as we have seen, observed that the sphex wasp 'lacked intelligence', and the wasp provides an even nicer example of tropistic behaviour. Wooldridge describes the same bit of wasp behaviour as Darwin, but in greater detail:

When the time comes for egg laying, the wasp *Sphex* builds a narrow burrow for the purpose and seeks out a cricket which she stings in such a way as to paralyze but not kill it. She drags the cricket into the burrow, lays her eggs alongside, closes the burrow, then flies away, never to return. In due course, the eggs hatch and the wasp grubs feed off the paralyzed cricket, which has not decayed, having been kept in the wasp equivalent of deep freeze. To the human mind, such an elaborately organized and seemingly purposeful routine conveys a convincing flavour of logic and thoughtfulness—until more details are examined. For example, the wasp's routine is to bring the paralyzed cricket to the burrow, leave it on the threshold, go inside and see that all is well, emerge, and then drag the cricket in. If, while the wasp is inside making her preliminary inspection, the cricket is moved a few inches away, the wasp, on emerging from the burrow, will bring the cricket back to the threshold, but not inside, and will then repeat the preparatory procedure of entering the burrow to see that everything is all right. If again the cricket is removed a few inches while the wasp is inside, once again the wasp will move the cricket up to the threshold and reenter the burrow for a final check. The wasp never thinks of pulling the cricket straight in. On one occasion, this procedure was repeated forty times, always with the same result.

Wooldridge suggests that we understand the wasp's behaviour using concepts borrowed from computer science: the insect's actions are controlled by a program with only a limited number of subroutines, with each subroutine called into play by a specific stimulus. The cricket-at-a-distance is the stimulus that triggers the subroutine for dragging-to-the-threshold; the cricket-on-the-threshold is the stimulus that triggers the subroutine for checking-out-the-burrow; and so on for each item in the wasp's limited behavioural repertory. So long as there is no interference, everything goes smoothly and we have the appearance of intelligence. But when the human hand interferes, the true nature of the performance is revealed: the wasp's behavioural program loops back to the earlier stage and runs mechanically forward again, oblivious to what has gone before.

What are we to make of this? One key question is whether all animal behaviour is to be regarded as tropistic, or only some of it. If the latter, then obviously there is no threat to Darwinian ideas. Darwin himself argued that the sphex is not intelligent, because it is unable to adjust its behaviour to changing circumstances. In Wooldridge's report Darwin would simply find confirmation of his own view; and if bees and moths are also unintelligent, what of it? So long as there is a continuum, with various animals possessing varying degrees if intelligence, it doesn't matter if some of the lower animals have none at all.

But what if *all* behaviour is tropistic, including the most impressive performances of the higher animals? It is hard to see why this could not be so. Consider: the wasp's behavioural repertory consists of stored subroutines that are triggered by specific stimuli. It is fairly easy to expose the tropism because the wasp's repertory is so small; it does not have a subroutine, for example, to specify that only one checking-out-the-burrow is necessary. But let us imagine that a subroutine is added to take care of this problem. Then the wasp's behaviour will seem a little more intelligent. We may go further, and imagine that many other subroutines are added, giving the wasp a much larger repertory of behaviours that respond to a much greater range of stimuli. The wasp might then appear to be *very* intelligent, as intelligent as a monkey or an ape. But, as Daniel Dennett has pointed out, we can never add enough subroutines to enable the wasp to cope intelligently with every possible situation:

There will always be room for yet one more set of conditions in which the rigidly mechanical working out of response will be unmasked, however long we spend improving the system. Long after the wasp's behaviour has become so perspicacious that we would not think of calling it tropistic, the fundamental nature of the systems controlling it will not have changed; it will just be more complex. In this sense any behaviour controlled by a finite system must be tropistic.

Since all animal behaviour is controlled by finite systems, we might by this reasoning come to regard even the 'intelligent' behaviour of the higher animals as in principle similar to the behaviour of the bee, the moth, and the wasp.

If one takes this view of animal behaviour, *and at the same time regards human behaviour differently*, then Darwin's opinion about the psychological similarities between humans and non-humans must be rejected. But once again, why should we regard humans differently? If we are willing to regard even the most complex behaviour of monkeys and apes as in principle tropistic, there is no reason not to think of humans in the same way. Human behaviour, too, is under the control of a finite system—the human brain—and this means that the human behavioural repertory, no matter how vast, also has its limits.

Many philosophers have found such a view of human capacities to be unacceptable; they have maintained that human behaviour is, in at least some respects, infinitely flexible. The favourite example of this is human language. Descartes thought that there are no limits to man's linguistic capabilities: there are an infinite number of possible sentences, and, Descartes added with a flourish, 'Even the lowest type of man can reply appropriately to *everything* that may be said in his presence.' And so Descartes thought that man's intellectual performances could not be under the exclusive control of a finite thing like the brain. It is worth noting, though, that this conclusion, however agreeable it is to our self-esteem, is not mandated by anything that we actually observe about human beings. Taken literally, Descartes's statement is false, because people are often stymied about what to say.

Human behaviour can be seen to be tremendously flexible, but not infinitely so. On the contrary, there is good reason to doubt that the human intellect has unlimited flexibility. Consider Descartes's own example of language-comprehension. We might say that there are, in principle, an infinite number of sentences possible in English, but only because there is no limit to the length of such sentences. English has a finite vocabulary. We could, therefore, formulate all possible two-word sentences; then we could formulate all three-word sentences; and so on. This process could in principle go on forever. However, we would eventually come to sentences that no human being could comprehend, because no one can grasp the import of a syntactically complicated million-word sentence.

Nevertheless, an individual's behaviour does not have to be infinitely flexible for it to be reasonable to regard its performances as rational. That is too high a standard. A lesser degree of flexibility will do. The key pattern is this. Suppose an individual wants X, and is able to adopt an effective strategy for getting X. But then circumstances change, and that strategy no longer works. Suppose the individual is then able to devise a new strategy, taking the new circumstances into account. If he is able to cope with a sufficient number of such changed circumstances, then his behaviour can reasonably be called rational. If he is unable to cope with some specific circumstances, then there is a failure of rationality at that point; but this does not mean that we must stop calling his other performances rational. This is one of the chief ways that rationality can be shown; and it is a pattern exemplified in the behaviour of many non-humans as well as humans.

ARE HUMANS THE ONLY MORAL ANIMALS?

'Of all the differences between man and the lower animals,' Darwin wrote, 'the moral sense is by far the most important.' As if to underscore its importance, he devoted a long chapter in *The Descent of Man* to a discussion of the nature and origins of morality. As one might have expected, Darwin argued that non-human animals have the same capacities that form the basis of morality in humans; although in non-humans, he observed, those capacities are not so well developed.

Morality is made possible, on Darwin's view, by our 'social instincts'—our natural disposition to act for the benefit of others. 'The moral sense', he wrote, 'is fundamentally identical with the social instincts.' The social instincts lead us to set aside our own narrow interests, and do what is for the good of the whole community. But other animals also have social instincts and are capable of acting self-sacrificially for the benefit of their fellow creatures. Therefore they should also be thought of as acting morally.

Darwin's argument follows the same general path as his argument about rationality: first, he gives a large number of examples of animal behaviour, designed to overwhelm the reader with evidence that animals do in fact have instincts that operate for the good of the community. Then he argues that this not surprising, but is precisely what one would expect on the hypothesis of natural selection. Along the way he protests the inconsistency of our unwillingness to attribute morality to the animals, when we would so characterize men for essentially the same behaviour. And finally, he offers an account of how the distinctive features of human morality could have evolved from non-human beginnings.

The 'Social Instincts' in Non-human Animals

Darwin begins by mentioning some simple instances of social instincts and then gradually proceeds to more impressive examples. The simplest evidence of sociability in animals is their living in groups and their apparent need for one another's company. 'Everyone must have noticed how miserable horses, dogs, sheep, etc. are when separated from their companions; and what affection at least the two former kinds show on their reunion.' But they do not merely live in proximity to one another: we may also notice that 'Social animals perform many little services for each other: horses nibble, and cows lick each other, on any spot which itches: monkeys search for each other's external parasites.'

The removal of external parasites is no small matter for an animal, but we are apt to be unimpressed by such examples. After all, it is easy to explain such patterns of behaviour as prompted by simple self-interest, as a matter of 'you scratch my back and I'll scratch yours', unrelated to anything that deserves to be called moral. But animals can be seen to perform even more valuable services for one another: mothers tenderly care for their babies; orphans are 'adopted' by other members of the group; and we even find instances of animals caring patiently for old or feeble companions. If we were searching for a clear example of moral behaviour among humans, we might choose a case of someone caring for an elderly invalid, without hope of reward. Perhaps with this in mind, Darwin quotes these reports:

Capt Stansbury found on a salt lake in Utah an old and completely blind pelican, which was very fat, and must have been long and well fed by his companions. Mr Byth, as he informs me, saw Indian crows feeding two or three of their companions which were blind.

Animals also warn each other of danger, and will even expose themselves to danger when it is necessary to rescue another from peril:

Brehm encountered in Abyssinia a great troup of baboons which were crossing a valley: some had already ascended the opposite mountain, and some were still in the valley: the latter were attacked by the dogs, but the old males immediately hurried down from the rocks, and with mouths widely opened roared so fearfully, that the dogs precipitately retreated. They were again encouraged to the attack; but by this time all the baboons had reascended the heights, excepting a young one, about six months old, who, loudly calling for aid, climbed on a block of rock and was surrounded. Now one of the largest males, a true hero, came down again from the mountain, slowly went to the young one, coaxed him, and triumphantly led him away—the dogs being too much astonished to make an attack.

Darwin refers to the baboon in this story as 'a true hero', and he has little apparent reluctance in referring to other animals as kindly, generous, and selfless. Yet, at the same time that these examples illustrate Darwin's optimistic estimate of animal 'morality', they also illustrate the paucity of evidence available to him concerning animal behaviour. Lacking reliable and detailed ethological studies, he had to depend on his own limited observations, on folk-wisdom, and on the hear-say reports of amateur observers. One of the striking things about Darwin's writing is the apparent confidence he felt in citing such anecdotal reports. It is worth remembering how much natural history was, in the nineteenth century, the work of lady and gentleman amateurs.

Nevertheless, subsequent ethological and psychological studies have confirmed Darwin's impression. Today we have a wealth of information about animal behaviour, garnered both in the field and in the laboratory, and these studies show that non-human behaviour is, if anything, even more 'human' than Darwin could have imagined. Let us look at one such study in some detail, remembering that it is only one from among many that might have been chosen.

Altruism in Rhesus Monkeys

Altruism might be defined simply as action that is motivated by the desire to help others. However, we may also use the word in a somewhat stronger sense, as involving *the willingness to forego some good for oneself* in order to help others. Understood in this stronger way, altruism is often taken to be the paradigmatic moral trait. But is altruism, in this sense, exclusively a human characteristic? Or do other animals also possess this quality?

To approach this question, let us examine a series of experiments conducted at the Northwestern University Medical School and reported in the psychological journals for 1964. These experiments were designed to discover whether rhesus monkeys are altruistic, and the method was to see whether they would be deterred from operating a device for securing food if doing so would cause pain to another monkey. One animal (called by the experimenters the 'operator' or 'O') was placed in one side of a divided box and taught to obtain food by pulling either of two chains. Food was available only when a light signal was given (a different light for each chain), and the O was trained to show no special preference for either chain.

Next, another monkey (called the 'stimulus animal' or 'SA') was put into the other side of the box, which was divided by a one-way mirror so that the O could see the SA but not the other way around. The floor on the SA's side was covered with a grid attached to a shock source. Three days were allowed for the O to adapt to the presence of the SA, and then a circuit was completed so that whenever the O pulled one of the chains to secure food the SA received a severe electrical shock. Pulling the other chain continued to give food, but produced no shock. Now, by turning on one signal light at a time, in various sequences and at various intervals, the experimenters could determine the extent to which the perception of the SA's distress would influence the O's willingness to pull the shock-producing chain.

After numerous trials the experimenters concluded that 'a majority of rhesus monkeys will consistently suffer hunger rather than secure food at the expense of electroshock to a conspecific'. In particular, in one series of tests, 6 of 8 animals showed this type of sacrificial behaviour; in a second series, 6 of 10; and in a third, 13 of 15. One of the monkeys refrained from pulling either chain for 12 days, and another for 5 days, after witnessing shock to the SA—which means they had no food at all during that time.

These experiments seem to show the rhesus monkeys are altruistic in the strong sense—that they will sacrifice their own good for the sake of others. The experimenters themselves reach this conclusion; however, they put the word 'altruism' in scare-quotes, apparently to indicate reservations about using it. But why the reservations? One reason might be that, lacking human language, the animals are not able to form abstract moral conceptions—they cannot think of themselves as altruistic, nor can they formulate the idea that altruism is a good thing. But, even for humans, *being* altruistic does not necessarily involve forming the idea that it is good to be altruistic, or that one is morally required to act altruistically. Being altruistic only requires desiring that others not suffer, and acting on that desire, even at cost to oneself. Animals may not form abstract conceptions, but they do have desires, and apparently it is a powerful desire of the rhesus monkey that he should not cause suffering to others of his own kind.

One might also be wary of ascribing altruistic motives to the monkeys because there are other possible interpretations of their behaviour. The monkeys' actions can be interpreted in various ways; how do we know which interpretation is right? Fortunately (for our curiosity, if not for the monkeys' welfare) the experimenters provided the additional information we need to answer this question.

First, they ran tests to determine whether the O's reluctance to pull the shock chain was correlated with relative positions in a dominance–submissiveness hierarchy. Relative dominance was determined when the animals 'were paired against each other in another apparatus and required to compete for 100 grapes presented one at a time. In most cases dominance was quickly established, the dominant animal getting 90 percent of the grapes.' The experiments were then divided into those in which the dominant animal was the SA and the submissive animal was the O; those in which the roles were reversed; and so forth. And it was found that this made no difference to the outcome.

Again, the experimenters were careful to observe whether differences in sex had any effect; that is, whether the O was male and the SA female; whether they were both male; and so forth. This made no difference either.

These results are important because they are exactly what we would expect if the O's behaviour is caused by a generalized altruism directed towards other members of his own species rather than to a fear of dominant animals or to some sort of gender-related impulse. The experimenters also rule out 'increased noise level' as a possible explanation because 'the SAs vocalized infrequently'—although, even if the SAs had cried out often, this would not rule out compassion as an explanation because the cries would so obviously be cries of pain. Moreover, the experimenters observe that 'the rage and attack mimetics of large male or female SAs during shock proved to be no more effective than those of smaller animals in deterring the feeding responses of persistently indifferent Os or expediting "altruism" in the others.' So still another alternative explanation is ruled out.

Other aspects of the experiments support the ascription of altruism in a different way. The experimenters found that animals who had previously been SAs were significantly more reluctant to pull the shock chain when they were made Os than animals who had not been SAs themselves. 'This behaviour of the shocked Os was not attributable to an acquired aversion to the apparatus itself since they showed no decrement in chain-manipulation during the adaptation sessions immediately following their shock.' The explanation suggested by the hypothesis of altruism is that these animals were more reluctant to pull the chain because, having suffered the shocks themselves, they had a more vivid comprehension of what it was like, and so a greater reluctance to see someone else in the same position.

It was also found that Os who had been cage-mates of their SAs were more reluctant to pull the shock chain than Os who had not been cage-mates of their SAs. Again, this is just what we would expect if we take our common knowledge of human beings as our model: we are less willing to harm someone we know than we are to harm strangers.

Taken together, these results provide impressive support for the view that rhesus monkeys are altruistic, by ruling out other possible interpretations of their behaviour, and by showing that their behaviour is influenced by factors similar to those that shape altruistic behaviour in humans. Still, many people are reluctant to attribute such moral qualities as altruism to mere animals, and we can anticipate various objections.

Someone might say: 'But the monkeys only showed an aversion to causing pain for others *of their own kind.* Would they do the same for *other* kinds of animals? The results here are much too limited to justify talk about a virtue of compassion.' The obvious answer to this is that even if the compassion of monkeys is limited to a feeling for others of their own kind, their compassion is no more limited than that of most humans. Most people—even those who have a fine respect for the interests of other humans—are fairly indifferent to the interests of beings not of their own kind. (For example, humans give little thought to arranging experiments in which electroshocks will be administered to members of other species.) We would not say that a human was completely devoid of compassion merely because he limits his concern to the suffering of other humans; and it does not seem right to judge other animals more harshly than we would judge ourselves.

Or someone else: 'But only *some* of the animals tested showed "altruistic" behaviour. Many did not.' Again, this is exactly the result we would get in the case of humans. Human compassion comes in varying degrees and strengths: some of us are quite compassionate, and some of us are relatively indifferent to the plights of others. And even those of us who do behave altruistically as a general rule may fail to do so on particular occasions. When we find similar variations among the monkeys, why should we be surprised? Indeed, the remarkable thing is that the differences are not greater than they are.

Explaining the Social Instincts

Before Darwin, our understanding of the nature of non-humans was controlled by a certain picture of the world: according to this picture, the gap between human nature and animal nature was established once and for all by God in his original act of creation. To men he gave souls, free will, rationality, and moral judgement; the other animals he created as lesser beings. Against the background of this picture, any attribution of moral qualities to animals would seem impossible. What is needed, in order to make such attributions possible, is the substitution of a different picture. Darwin provided the new picture, and tried to show that once it is adopted the view of animals as (at least partially) moral beings follows naturally.

Darwin believed that the existence of the social instincts could be explained as the result of natural selection. The key, as always, is to understand how individuals who possess this characteristic are better situated in the struggle for survival. This is not easy to understand; in fact, where altruism is concerned, just the opposite seems true: the tendency to behave altruistically seems to work *against* reproductive success. Consider Darwin's own example of the heroic baboon who risks his life to rescue his young companion. Such behaviour might increase the chances of the *youngster's* surviving to reproduce, but it obviously decreases the chances of the hero's survival. Thus the characteristics that will be passed on to future generations will most likely not be the rescuer's characteristics, but the characteristics of the one rescued. How, then, is the tendency to rescue supposed to be preserved?

This question would be easy to answer if we imagined that 'the struggle for survival' is a competition between species rather than individuals. We could then point out that, if the members of a group—whether it is a local group or an entire species—are helpful to one another, that group is more likely to survive than a group whose members do not aid one another. This sounds tempting, but it cannot solve our problem, for it does not explain how altruism can be established *within* the group. How does altruism become a widespread characteristic within a group in the first place? Considering that individual altruists such as our heroic baboon seem to be at a disadvantage, why shouldn't the tendency to altruism be eliminated the moment it first appears? In order for natural selection to favour a tendency to rescue, there must be some advantage *for the rescuer*, and not merely for those he rescues, who, after all, might not share his generous inclinations.

Darwin was well aware of this problem. He wrote:

But it may be asked, how within the limits of the same tribe did a large number of members first become endowed with these social and moral qualities, and how was the standard of excellence raised? It is extremely doubtful whether the offspring of the more sympathetic and benevolent parents, or of those which were the most faithful to their comrades, would be reared in greater number than the children of selfish and treacherous parents of the same tribe. He who was ready to sacrifice his life, as many a savage has been, rather that betray his comrades, would often leave no offspring to inherit his noble nature. The bravest men, who were always willing to come to the front in war, and who freely risked their lives for others, would on an average perish in larger numbers than other men. Therefore it seems scarcely possible (bearing in mind that we are not here speaking of one tribe being victorious over another) that the number of men gifted with such virtues, or that the standard of their excellence, could be increased through natural selection, that is, by the survival of the fittest.

Darwin's statement of the problem is far more impressive than his somewhat half-hearted attempt at a solution. Apparently feeling that he had to say *something* about the evolution of 'the social instincts', he speculated that intelligent creatures would learn from experience that helping others brings help in return, and that helping behaviour, once begun, would become habitual and would be reinforced by social praise and blame. It seems clear, though, that such remarks leave the deep problem untouched.

The fact is that Darwin did not know how to explain the competitive advantage conferred by the social instincts, and after a little humming and hawing he was candid enough to say so. Throughout nature, the most common, powerful, and steadfast altruism is that shown by parents for children and siblings for one another. What accounts for their behaviour? Darwin was mystified:

> With respect to the origin of the parental and filial affections, which apparently lie at the basis of the social affections, it is hopeless to speculate; but we may infer that they have been to a large extent gained through natural selection.

The 'inference' of which Darwin speaks here is more a matter of faith in his theory than anything else: he says that 'we may infer that they have been to a large extent gained through natural selection' because he is confident that *everything* is to a large extent gained through natural selection. But how, exactly, are *these* affections 'gained through natural selection'? That is what Darwin did not know. Today, however, we have some idea of how this works, because we understand better than Darwin did the mechanisms by which natural selection operates. Darwin did not know about genes, and we do.

Genes, of course, are the biological units that determine some (but not all) of an individual's characteristics. The reason children tend to resemble their parents is that, in the course of reproduction, parents pass on their genes to the children. A human child will inherit half his father's genes and half his mother's genes. With this in mind, we may redefine what it means for a characteristic to confer an advantage in the 'struggle for survival': a characteristic confers an advantage if it increases the likelihood that one's genes will be preserved in future generations.

To see how this works, suppose there are two animals, Loving Mother and Indifferent Mother. (You can think of them as baboons, kangaroos, pigs, humans, or anything else—it doesn't matter.) Loving Mother has genes that dispose her to be protective of her offspring; Indifferent Mother does not. Therefore Loving Mother will try to ensure her children's welfare, while Indifferent Mother will show no special concern for them. Whose genes are more likely to be represented in future generations? Plainly, Loving Mother's genes have the better chance. Her genes are more likely to survive because the carriers of those genes—her children—are more likely to survive. Thus, as the generations pass, the genes that dispose individuals to behave protectively towards their children will tend to spread throughout the population, while the genes that permit indifference will tend to disappear.

But it is not only one's children who share one's genes. So do one's brothers and sisters. Again, think of two animals, Loving Sibling and Indifferent Sibling. Loving Sibling has genes that dispose her to help her brothers and sisters; Indifferent Sibling does not. Because Loving Sibling's brothers and sisters have a helper, they are more likely to survive to reproduce, and so *those* genes are also more likely to be passed on to future generations.

In this way we may account for the simplest and most powerful cases of altruistic behaviour, which Darwin called 'the parental and filial affections'. Viewed in this light, there is nothing mysterious about the self-sacrificial altruism shown by parents and siblings. It is no more than we should expect, given how natural selection operates. The point is not that individuals *calculate* how to ensure the survival of their genes—no one does that. The point is that these are types of genetically-influenced behaviour that will be preserved by the same mechanism that preserves the polar bear's warm coat, the finch's well-shaped beak, and any other advantageous characteristic.

'Kin altruism', as it is called, leads individuals to care for their relatives just to the extent that those relatives share the individual's genes. This explains why we are especially concerned for the welfare of our children and siblings, somewhat less for our cousins (who share fewer of our genes), and even less for strangers. Sociobiologists have found ample evidence that this is exactly how helping-behaviour works in the animal world. Here is a remarkable example, from the work of R. L. Trivers and H. Hare:

Most sexually reproducing animals get half their genes from each of two parents; therefore, siblings will, on average, share half of each other's genes. We may say that such siblings are related by the ratio 1/2. But there are exceptions to this. Ants reproduce in such a way that females are related to their brothers by 1/4 and to their sisters by 3/4. Thus a female ant shares three times as many genes with her sisters as with her brothers.

This means that, if the inclination to help others varies with the degree of kinship, we should expect female ants to be three times more concerned for the welfare of their sisters than their brothers. As it turns out, there is a convenient way to determine how much assistance a female ant will give to her siblings. In the ant colony workers provide food for the others; so we can measure the female worker's preference for her brothers or sisters by observing the amount of food she provides them. Trivers and Hare studied these feeding patterns, and found that almost exactly three times more food was being provided for sisters than for brothers! Moreover, a study was made of comparable behaviour on the part of workers who had been 'enslaved' and made to work for queens unrelated to them. When this happened, the amount of food provided for males and females became almost equal.

Thus we have found an explanation for 'the origin of the parental and filial affections' about which Darwin thought it was 'hopeless to speculate'. The genes that lead one to care for one's relatives will be preserved by natural selection in the same way that any other genes are preserved.

But not all altruism is kin altruism. Animals may be observed to sacrifice their own interests to help others who are not closely related to them, and this is more difficult to explain. In an attempt to deal with this problem, sociobiologists have introduced a concept of 'reciprocal altruism'. The notion of reciprocal altruism is, however, more problematic than that of kin altruism. The basic idea is that an individual performs a service for another because doing so increases the likelihood that a similar service will be performed for him—a monkey picks the external parasites off the back of another monkey, and then the favour is returned. It is easy enough to see that such reciprocal aid, when practised by all (or even most) members of a group, will work to the advantage of all; but it is not so easy to see how, on the principles of natural selection, such behaviour could become established in the first place.

There is, at this time, no perfectly satisfactory solution to this problem. The explanation of reciprocal altruism remains murkier than that of kin altruism. But we may make an educated guess about the direction from which a solution might come. It is at this point that Darwin's own explanation of altruism might have some value. Darwin may have been on the right track when he associated the higher forms of social affection with reason: speaking of the development of altruism in man, he said,

In the first place, as the reasoning powers and foresight of the members became improved, each man would soon learn from experience that if he aided his fellow-men, he would commonly receive aid in return. From this low motive he might acquire the habit of aiding his fellows; and the habit of performing benevolent actions certainly strengthens the feeling of sympathy, which gives the first impulse to benevolent actions.

The key idea here is that individuals' cognitive powers might play an important part in the establishment of reciprocal altruism within a population. This would make the explanation of reciprocal altruism importantly different from the explanation of kin altruism, which, as we have seen, requires no such assumptions about reasoning.

Following Darwin's suggestion, we might envision a process that takes place in three stages. In the first stage, there is only kin-altruism: individuals aid their relatives, but no one else. The 'social instincts' are in place, but they do not govern behaviour beyond one's kin. In the second stage, individuals attain enough 'reason and foresight' to understand that aiding non-relatives might be a good strategy for gaining some benefit for themselves, provided that the non-relatives can be induced to reciprocate. This might at first be a simple thing: A and B both have parasites that are hard to reach, and they both want them removed; in casting about for a way to accomplish this, A removes B's parasite and then presents himself, in a suggestive posture, to B; B 'catches on' to the offer, and sees that his own welfare is being served by playing this game of tit-for-tat, so B then removes A's parasite. Thus the social instincts are extended beyond one's kin. In the third stage, this pattern of behaviour becomes more widespread, until it has become habitual.

Is this anything more than a suggestive fantasy? Why should we think that the key to understanding altruism beyond one's relatives might be something like this? It is significant that all the most impressive examples of non-kin altruism are from the so-called 'higher' animals—humans, monkeys, baboons, and so on—animals in which the power of reasoning is well developed. In the 'lower' animals we find only kin altruism. This seems to confirm Darwin's speculation that the development of general altruism might go hand-in-hand with the development of intelligence. The hypothesis would be that animals are capable of altruism towards non-relatives only to the extent that they are intelligent enough to form beliefs about whether their aid is likely to be reciprocated The hypothesis is not that the non-humans must be able to *articulate* such beliefs; it is enough that they be able to form expectations about one another's behaviour, and adjust their own behaviour accordingly.

It is also important to remember that we are only trying to account for a limited phenomenon. It we start with the assumption that humans exhibit a kind of grand, Sermon-on-the-Mount altruism, and we then assume we are trying to explain *that*, then Darwin's suggestion might seem altogether too feeble. But we should be careful not to overstate the extent of non-kin altruistic behaviour. By far the most powerful kind of altruism, even among humans, is kin altruism. Even when people do show an unselfish willingness to help strangers, their preference for helping their own kin remains very much stronger: our non-kin altruism is so weak that when an affluent American gives a few hundred dollars to support famine-relief efforts, while spending thousands to send his children to an expensive university, he is judged to be exceptionally generous. Truly disinterested, generalized saintliness might exist in a few people, but it is so rare that it may be regarded, in the naturalist's terms, as a mere 'variation'—and whether it is something that *could* spread to the population as a whole might well be doubted.

The Special Moral Capacities of Human Beings

Despite his concern to demonstrate continuities between human and animal life, Darwin does not argue that non-humans are moral agents in the same sense as humans. Here, as everywhere, his views are tempered by a vigorous common sense. There are obviously differences between the moral capacities of humans and non-humans, and Darwin does not deny it. Instead, he sets himself to explain what those differences are and why they exist.

In his discussion of human morality, Darwin is more explicitly philosophical than in other parts of his work. He offers a definition of morality, a conception of how moral knowledge is acquired, an account of the nature of the human conscience, a description of what makes a man good or bad, and an estimate of the path that must be taken if moral progress is to be made. All this seems, at first glance, to be far from the proper concern of the biologist. Darwin realizes that he is covering ground that is traditionally reserved for the moral philosopher, but he thinks it is nevertheless worthwhile to consider how these matters might appear when viewed 'from the side of natural history'.

What morality is. Darwin thinks that, when ethics is viewed from the perspective of natural history, we find reason to reject the opinion of some philosophers that 'the foundation of morality lay in a form of Selfishness'. It is the social instinct, and not the instinct for self-preservation, that forms its basis. Therefore, Darwin says, those philosophers who have advocated 'the Greatest Happiness principle' are closer to the mark—although he thinks it would be better to speak of 'the general good or welfare of the community, rather than the general happiness'. (The latter, in his opinion, is too restricted a term.) Of course, Darwin believed that the social instincts arise only because they confer an advantage in the struggle for survival; and so in some sense the drive for self-preservation is more fundamental. But *morality* comes into existence only with the social instincts. Hence Darwin's insistence that 'the Greatest Happiness principle' more aptly expresses the basic moral rule.

An unsympathetic reader might easily find in this a violation of Hume's stricture against deriving 'ought' from 'is'. Consider this passage, in which Darwin suggests a 'definition' of morality that encompasses both human and non-human conduct:

As the social instincts both of man and the lower animals have no doubt been developed by the same steps, it would be advisable, if found practicable, to use the same definition in both cases, and to take as the test of morality, the general good or welfare of the community.

Darwin seems to assume that a particular conception of right and wrong ('We ought to do whatever promotes the general good or welfare of the community') follows from the facts about how our social instincts have developed. But this is just the fallacy that Hume warned us about. As a naturalist, one might say, Darwin is entitled to describe the historical development of human capacities, including behavioural capacities, and to theorize about their underlying causes. Logically speaking, however, nothing follows from this about what is good or evil: it remains an open question whether the historical process has resulted in good or bad tendencies of action. And what is more (the critic might continue), Darwin seems to have naïvely assumed the truth of a particular moral philosophy, utilitarianism, without realizing how controversial and open to objection that moral philosophy is.

But this would be too unsympathetic a reading. Darwin is not trying to prove what is right and what is wrong; nor is he trying to promote a contentious philosophical theory. He begins by formulating a conception of what morality is because he has no choice. If one is trying to understand the development of the moral capacities, one must start with some idea of what the moral capacities are; otherwise one might end up explaining the development of something entirely different. This creates trouble at the beginning, for any conception of the moral capacities is bound to be controversial—whatever understanding of right and wrong one proposes, no matter how innocuous, there is some philosopher who will dispute it.

At any rate, the understanding of good and bad conduct that Darwin brings to his discussion is not morally eccentric: it is about as innocuous as any such conception could be. Appearances aside, it is more or less neutral between the main competing moral philosophies. It is broad enough to be compatible with the basic ideas of both utilitarian and Kantian conceptions. Darwin assumes that moral behaviour promotes the general welfare, but he also stresses that a moral agent is an individual with a conscience—a sense of duty—not unlike that envisioned by Kant. The two notions are wedded by Darwin's assumption that a person of conscience will standardly approve of behaviour that promotes the general welfare. The task he sets himself is to explain, compatibly with the principles of natural selection, how humans could have come to be moral agents of this sort.

Reason and the acquisition of moral knowledge. Human morality is the product, not just of the social instincts, but of the social instincts plus intelligence. Thus, on Darwin's view, the higher stages of morality emerge with the development of the higher rational capacities: 'Any animal whatever, endowed with well-marked social instincts, would inevitably acquire a moral sense or conscience, as soon as its intellectual powers had become as well developed, or nearly as well developed, as in man.'

But what difference does intelligence make? The answer is that behaviour, at least in the higher animals, is the combined product of attitudes and beliefs. In the simplest instance, one desires something, and believes that one can get it by acting in a certain way. Obviously, the less one knows about causes and effects, the less one will be able to choose actions that lead to what one wants. The social instincts may be understood as a set of attitudes, namely, the attitudes that consist in desiring the general welfare. Therefore, the less intelligent an animal is, the less well equipped it will be to choose actions that satisfy *that* attitude. Darwin notes that, as primitive men accumulate a rude sort of knowledge, they make mistakes, and this causes them to adopt defective rules of conduct—defective in the sense that the rules seem reasonable only because of the cognitive errors. 'Savages, for instance, fail to trace the multiplied evils consequent on a want of temperance, chastity, etc.' But as they become more sophisticated intellectually, they will be better able to trace those evils, and their conduct will improve. In this way moral knowledge grows.

The social instincts which no doubt were acquired by man, as by the lower animals, for the good of the community, will from the first have given to him some wish to aid his fellows, and some feeling of sympathy. Such impulses will have served him at a very early period as a rude rule of right and wrong. But as man gradually advanced in intellectual power and was enabled to trace the more remote consequences of his actions; as he acquired sufficient knowledge to reject baneful customs and superstitions . . . so would the standard of his morality rise higher and higher.

Conscience. Eighteenth-century Britain produced a distinguished group of moral philosophers, including such figures as Lord Shaftesbury, Joseph Butler, and David Hume. The idea of a 'moral sense', or conscience, was prominent in their thinking, but they gave the notion a decidedly naturalistic interpretation. To have a moral sense was to have a capacity for second-order attitudes—attitudes that have one's *other* attitudes as their objects. This, they thought, was what makes man a moral agent in a sense in which other animals are not. A dog's attitudes (they said) are all directed at objects external to the dog himself: he desires food, he desires what will make him warm, he desires to avoid the sources of pain. Perhaps, they might have said if they had known more about altruism among the animals, a dog might even desire that other dogs should not suffer. But the dog cannot desire to have a certain attitude, and he cannot regret that he has certain attitudes. A man, on the other hand, can want something (I want to hurt the person who hurt me) and at the same time can regret that he wants it (I disapprove of myself for wanting revenge, and wish that I had a more generous temperament). It is this capacity for approving or disapproving of one's own attitudes that constitutes one's conscience.

In Darwin's notebooks for 1838–40 there are numerous references to these thinkers, and Darwin's own treatment of conscience, elaborated years later in *The Descent of Man*, is similar to their treatment. But Darwin thought that the very existence of conscience is puzzling: How can there be such a thing? First let me explain the puzzle, and then we will look at Darwin's solution to it.

Conscience, as Darwin conceives it, is a phenomenon closely associated with *conflict situations*. Facing a choice about what to do, a man may be pulled in different directions by conflicting 'natural impulses'. For example, there may be a danger that he can avoid by running away, and he may be afraid; or, there may be food that he can seize, and he may be hungry. At the same time, running away from the danger or seizing the food may be contrary to the interests of the community at large, and because he has 'social instincts' he is disinclined to do those things. So whatever he does, he will be going against one or another of his natural impulses. Suppose, then, his fear or his hunger wins out, and he acts contrary to the general welfare of the community. Later, when he reflects on what he has done, he feels bad about this, and regrets it. This is the deliverance of conscience; it is the wish that one, rather than another, of one's attitudes had prevailed.

Darwin found this puzzling because it seemed strange that one sort of 'natural impulse'—the social instincts—should be thought better or more worthy of respect than any other. Why, in retrospect, are we uncomfortable with the fact that we acted on some instincts rather than others, when they are all equally 'natural'? As Darwin put it,

Why does man regret, even though he may endeavour to banish any such regret, that he has followed the one natural impulse, rather than the other; and why does he further feel that he ought to regret his conduct? Man in this respect differs profoundly from the lower animals.

The eighteenth-century moralists had considered two general solutions to this problem. First, we can distinguish between the *strengths* of the various impulses to action. Perhaps the social instincts are simply stronger than the others, so that when the man regrets giving in to fear or hunger the regret is produced by the superior strength of the social impulse. This sort of explanation, however, does not seem to work. If the social impulse is stronger than fear or hunger, then why he did not act on it originally?

This sort of consideration led Butler—an Anglican bishop whose works were standard university fare during Darwin's day—to distinguish the *strength* of conscience from its *authority*. He held that the various springs of action form a natural hierarchy, with conscience at the top, so that conscience has a supremacy that does not depend on its strength. Thus, even if an individual constantly acts contrary to conscience because his conscience is weak, while his other impulses are strong, his conscience still dictates what he *ought* to do because that is its natural function. Darwin, however, regarded this as so much hocus-pocus, and sought a less exotic explanation.

'Butler and MacIntosh', wrote Darwin in his notebooks, 'characterize the moral sense by its "supremacy"—I make its supremacy solely due to greater duration of impression of social instincts, than other passions, or instincts.' It was this notion of 'greater duration' that was to provide the key for Darwin's explanation. In *The Descent of Man*, Darwin's account goes like this: suppose a man does something, out of fear or hunger, that harms the community. Then he reflects on what he has done. Now the social instincts are permanent, and persistent; but particular desires come and go. Therefore, when he reflects on his past conduct, his social instincts are still with him, but the particular desire that, at the time of action, overwhelmed the social instincts, is fading away. Thus he regrets what he did. This after-the-fact reflection is what we call 'conscience', and the fact that the social instincts are stronger at the time of reflection, even if they were not stronger at the time of the action, explains why such reflection results in their endorsement.

Thus, as man cannot prevent old impressions continually passing through his mind, he will be compelled to compare the weaker impressions of, for instance, past hunger, or of vengeance satisfied or danger avoided at the cost of other men, with the instinct of sympathy and good-will to his fellows, which is still present and ever in some degree active in his mind. He will then feel in his imagination that a stronger instinct has yielded to one which now seems comparatively weak ... At the moment of action, man will no doubt be apt to follow the stronger impulse; and though this may occasionally prompt him to the noblest deeds, it will far more commonly lead him to gratify his own desires at the expense of other men. But after their gratification, when past and weaker impressions are contrasted with the ever-enduring social instincts, retribution will surely come. Man will then feel dissatisfaction with himself, and will resolve with more or less force to act differently in the future. This is conscience; for conscience looks backwards and judges past actions, inducing that kind of dissatisfaction, which if weak we call regret, and if severe remorse.

The difference between good people and bad people. Having come this far, it should now be clear what the difference between good people and bad people is, on Darwin's view. A thoroughly admirable person will be one whose social instincts are strong enough to overcome the particular inclinations—fear, hunger, etc.—which might otherwise lead him to act contrary to the general welfare. (Although Darwin does not say so explicitly, we can explain easily enough why we regard this sort of person as admirable: it is because of the strength of *our* social instincts.) But not everyone is completely admirable, and there is still another distinction to be drawn between those who yield to temptation but later regret it, and those who have no regrets. It is the latter whom Darwin characterizes as 'essentially bad', for their social instincts are so weak, or even non-existent, that they do not even control later reflections. These are the people who, as we would say, have no consciences:

If he has no such sympathy, and if his desires leading to bad actions are at the time strong, and when recalled are not overmastered by the persistent social instincts, then he is essentially a bad man; and the sole restraining motive left is the fear of punishment.

How can there be such people? The natural explanation, from the standpoint of natural history, is that they are variations.

How moral progress is made. Despite his acknowledgement of the existence of conscienceless men, Darwin was an optimist who believed, in good nineteenth-century fashion, that the human race is advancing towards ever greater moral perfection. His account of the nature of this progress, and of the conditions of human life that make it possible, followed naturally from his view about the role of reason in expanding the scope of the social instincts. He envisioned an ever-widening circle of moral concern, that would spread beyond self and family to include neighbours and countrymen and eventually all mankind. The abolition of slavery, which he abhorred, would be the natural result of this expanded moral consciousness. In this Darwin anticipated the moral vision that has animated twentieth-century liberalism, with its emphasis on equal rights for all people. But he thought that the final extension of the human social instinct would go a step beyond even that, to include within its sphere of concern the other animals as well:

As man advances in civilisation, and small tribes are united into larger communities, the simplest reason would tell each individual that he ought to extend his social instincts, and sympathies to all the members of the same nation, though personally unknown to him. This point being once reached, there is only an artificial barrier to prevent his sympathies extending to the men of all nations and races. If, indeed, such men are separated from him by great differences in appearance or habits, experience unfortunately shows us how long it is before we look to them as our fellow-creatures. Sympathy beyond the confines of man, that is humanity to the lower animals, seems to be one of the latest moral acquisitions . . . This virtue [sympathy for the lower animals], one of the noblest with which man is endowed, seems to arise incidentally from our sympathies becoming more tender and more widely diffused, until they are extended to all sentient beings. As soon as this virtue is honoured and practised by some few men, it spreads through instruction and example to the young, and eventually through public opinion.

Again, Darwin says that, as man became more rational and as his social instincts were guided more and more by reason, 'his sympathies became more tender and widely diffused, so as to extend to the men of all races, to the imbecile, the maimed, and other useless members of society, and finally to the lower animals—so would the standard of his morality rise higher and higher'. It is an idealized version of history, at best. Whether this kind of progress is made inevitable by human nature may certainly be doubted. That this moral vision must be embraced by an adequate moral philosophy is, however, one of the main themes of my concluding chapter.

ANTHROPOMORPHISM AND THE RELEVANCE OF EVOLUTION

Darwin believed that animals have rational and moral capacities similar to our own. Others have resisted this conclusion, as he anticipated they would. Sometimes the resistance has been motivated by religious or philosophical dogma, but in many instances the opposition has come from scientists who contend that Darwin went too far in what he was willing to attribute to his beloved animals.

The debate about whether animals have such capacities, and what form they take, might proceed as a series of arguments about specific cases. Thus, someone who takes a generous view might produce an example of animal rationality (such as an octopus figuring out how to get food by screwing off the top of a glass jar) or of animal morality (such as the altruism of the rhesus monkey) and argue that such examples prove the point. This, however, only invites the sceptic to look more closely at the examples, and to suggest other, more modest ways of interpreting the animals' behaviour. Then, having impeached the examples, the sceptic is free to argue that extravagant attributions of reason and morality are not necessary and that they go beyond what the evidence really warrants.

Arguments about specific cases might seem endless. The outcome of the debate, however, does not depend entirely on how we choose to describe particular cases. There are some more general points that should be kept in mind, and these points provide good reason for thinking that, in at least some instances, Darwin's view must prevail.

The first such general point is this. In many cases, it may be possible to interpret an animal's behaviour as rational; but it may also be possible to explain the animal's behaviour in some other way, as the result of non-rational factors. The interpretation we are to place on the animal's behaviour—rational, or not?—may be underdetermined by the facts. In such cases, what are we to say?

If we were considering this question in the absence of a well-confirmed theory of origins, it might be reasonable to choose the latter option. After all, human intellectual capacities are so far superior to those of mere animals that it might well be doubted whether the animals have anything resembling our abilities. What have we to do with them? Before Darwin, this approach was almost universally taken.

Now, however, things are different: we have a well-confirmed theory which tells us that humans are closely related to other species. We know that we have evolved, over time, from the same ancestors as the monkeys and baboons. With such a theory in hand, it becomes more reasonable to see connections between human abilities and those of other animals. If it is unclear whether a particular bit of animal behaviour should be described as analogous to human behaviour, the fact that both originated in a common historical source provides *an extra reason* (and, one might well think, a compelling reason) in favour of interpreting them as analogous. This, I think, is extremely important. It means that the plausibility of viewing animals as rational (and to some degree moral) does not rest entirely on the outcome of individual arguments about how individual bits of animal behaviour should be construed.

A second and closely related point is this. Denying that other animals are rational involves positing a sharp break between humans and the members of other species—it is to say that we humans have characteristics that are found nowhere else in nature, not even in an attenuated form. Before Darwin, this could be reasonably believed. But, in the light of evolutionary theory, this would be altogether fantastic. Evolutionary theory leads us to expect continuities, not sharp breaks. It implies that, if we examine nature with an unbiased eye, we will find a complex pattern of resemblances as well as differences. We will find, in humans, traces of their evolutionary past, and in other species—especially those more closely related to us by lines of evolutionary descent—traces of characteristics that may be more or less well developed in us. This is true of those characteristics that make us 'rational', no less than the others.

Notwithstanding all this, it is only a contingent matter that there exist today other species that share particular human characteristics. It might have been otherwise. Suppose that, for some combination of reasons, all the other descendants of our near ancestors had died out, so that today there were no other mammals—no monkeys, no baboons, no dogs, etc. Then it would be much more plausible to think that man alone, among the current inhabitants of the world, is rational. The other animals that most obviously *might* have shared this characteristic would be gone. But of course this did not happen. It is because the current population of the world does include our near kin that it is unreasonable to suppose that humans will have major characteristics, including psychological characteristics, not found anywhere else.

Finally, if one is nevertheless tempted to believe that humans are psychologically unique, it is useful to remember that the whole enterprise of experimental psychology, as it is practised today, assumes otherwise. Animal behaviour is routinely studied with an eye to acquiring information that can then be applied to humans. Psychologists who want to investigate maternal behaviour, for example, but who are constrained by ethical considerations from experimenting with human mothers and infants, might study the behaviour of rhesus monkey mothers and infants, assuming that whatever is true of them will be true of humans—because, after all, they are so much like us. Harry Harlow, one of many psychologists who has specialized in just this sort of research, has written in its defence that rhesus monkeys not only love their offspring, and care for them as we do our own, but are highly intelligent, and 'can indeed solve many problems similar in type to the items used in standard tests of human intelligence'. (I will have more to say about this sort of research, and Dr Harlow's work in particular, in Chapter 5.) Those who continue to deny that animals have mental capacities similar to those of humans are implying that this research is fundamentally misguided. They thus join the long dismal parade of thinkers who have attempted to set aside whole fields of experimental science on *a priori* grounds.

The Charge of Anthropomorphism

The question of anthropomorphism hangs like a cloud over this whole discussion, and it is necessary, before concluding, to say something about it.

The charge is more often lodged against ethologists who study animal behaviour in the field than against experimental psychologists working in the laboratory; the reason, perhaps, is that the experimentalists tend to be dry and cautious in what they say. But when we turn to works such as Jane Goodall's study the chimpanzees of Gombe, or Dian Fossey's observations of the mountain gorillas of the Virungas, we find something entirely different. They offer absorbing accounts of complex behaviour that fully support Darwin's view of animals as intelligent, social, and even moral beings. However, their descriptions of the animals often seem intemperate and subjective, and so they are frequently dismissed as naïve enthusiasts who bring to their work a sentimental humanism that leads them to endow the animals with human 'personalities' all too easily.

Consider Fossey's description of the mountain gorillas. She starts by giving them cute human names—she calls her gorillas Puck, Beethoven, Peanuts, Augustus, Coco, Poppy, Lisa, Effie, Flossie, Pablo, and Uncle Bert—and then she seems intent on searching out analogies with human life and behaviour whenever she can. She describes them as peaceful animals who will not attack strangers unless threatened; but to keep peace with them, one must approach them in the right way, openly. Otherwise they will charge. These charges, frightening as they are, are said to be merely defensive. If one stands one's ground, and offers no resistance, they will back off. Fleeing, on the other hand, is a mistake:

A very capable student once made the same mistake as I had when approaching Group 8 from directly below. He was climbing through extremely dense foliage in a poacher area and noisily hacking at vegetation with his *panga*, not knowing the group was near. The faulty approach provoked a charge from the dominant silverback, who could not see who was coming. When the young man instinctively turned and ran, the male lunged toward the fleeing form. The gorilla knocked him down, tore into his knapsack, and was just beginning to sink his teeth into the student's arm when he recognized a familiar observer. The silverback immediately backed off, wearing what I was told was an 'apologetic facial expression' before scurrying back to the rest of Group 8 without even a backward glance.

So we have the silverback facing an unknown danger, exposing himself to protect his companions, but then recognizing a familiar face and reacting apologetically—an altogether reasonable, and even sensitive, performance. But the talk of 'apologetic facial expressions' is apt to provoke raised eyebrows even from those sympathetic to the idea of animal rationality.

Interestingly, scientists will sometimes try to have it both ways: they will produce one account, dry and cautious, for their academic peers, and another, more 'humanized' account for the general public. John Mackinnon, who studied orang-utans in Borneo, published a scholarly paper about his findings in the journal *Animal Behaviour*, and then wrote a popular book called *In Search of the Red Ape*. Behaviour that seemed remarkably human in the book is characterized in quite different terms in the paper. In the book we read that

Ruby's face grimaced with fear as Humphrey seized her from behind and dragged her out of her nest. The ardent lover bit and struck her then, clasping his feet firmly around her waist, proceeded to rape the unfortunate female.

Humphrey and Ruby—a rough but ardent lover rapes his woman. In the scholarly journal, however, the same scene plays quite differently:

Due to the small size of the male penis and the difficulty of suspended copulation, it is probable that only when the female cooperates in the mating can successful intromission be achieved. In one observed instance of 'rape' the female continued to struggle throughout and the male's penis could be seen thrusting on her back.

Not only are the names gone, and the word 'rape' placed inside scare-quotes, but we are now told that the attempted copulation was not even successful. Now it is merely a matter of an unco-operative female thwarting the male's attempt.

There are other instances of this phenomenon. Keith Laidler was one of those who tried to teach language to an orang-utan, and we have two accounts of his experience: one in a paper written for other investigators, and one in a popular book intended for a general audience, *The Talking Ape*. The account given in the book leads one to believe that the orang-utan's linguistic capacities were most impressive:

As time progressed, our little 'talking ape' began to use his simple functional language to insist on what he wanted even when it was out of sight. This began during Cody's 14th month of life during a feed session. Cody refused his last spoonful of cereal and instead uttered his 'milk' word. I hesitated and Cody repeated 'milk' in louder tones. I was intrigued as to what the young creature would do and placed him on the floor at my feet, whereupon Cody crawled to the kitchen for his cup of milk. Was he trying to tell me he wanted no more solid food while at the same time indicating what he preferred? Was he asking for his milk? I was not certain that he was, in fact, capable of such communicatory skill but a second incident convinced me.

Cody was playing on the floor when I approached with a pan of porridge and honey, a special treat the young orang really enjoyed. The following 'conversation' ensued:

Cody: 'Pick me up.'

Me: (having expected Cody to utter his usual 'food' sound) 'No. What do you say?'

Cody: (halfway up my trunk) 'Pick me up.'

Cody then reached the upper portions of my chest, and settled himself comfortably on my right hip. Only then did he turn and looking straight at the plate of porridge, say 'food.'

How can one doubt that apes can talk, when one reads accounts such as this? However, in the professional paper, one finds a more sobering report of the same incident:

Cody very rarely used his sounds to request some object presently out of sight, nor did he use one sound so as to put himself in a position to emit a second sound. This is considered due, in part, to the immaturity of the subject who was 15 months at termination of the experiment. Exceptions to this state were few, but did, nevertheless, occur, the protocol given below falling within the last week of the experiment and suggestive of a developmental, rather than an absolute, limitation on ability. Thus, on 15 October the infant twice refused the last of his pan-food voicing 'kuh' each time. When placed on the floor, he immediately made his way to where the milk bottle was located. Later after termination of the experiment, but before Cody had been left to return to a more natural state with a second infant orang, the infant came across to the teacher when he offered pan-food, voicing two 'puhs'. The teacher replied 'No, fuh' on two separate occasions (the correct sound for pan-food) but the infant replied each time with a 'puh,' attempting at the same time to climb up the teacher. When allowed to do so, and settled on the teacher's right hip, he turned and without being prompted, uttered a good fuh-sound with his eyes directed toward the food.

Now Cody's abilities seem much more modest.

What are we to make of all this? The sensible approach, it seems to me, would be to agree from the outset that anthropomorphism is a sin to be avoided, and to recognize that researchers have in fact often been so eager to find 'human' qualities in animals that they have distorted their data. The frequency of this problem gives one reason to be especially cautious before believing sensational claims. At the same time, it would be unreasonable to think that researchers are *always* guilty of this sin, whenever they find that animals have previously unsuspected qualities. Think again of the study of altruism in rhesus monkeys. In that case the attribution of altruism was supported, in considerable detail, by experimental results, and alternative hypotheses were carefully ruled out. So the evidence is that the altruism is really there—we are not merely 'reading it into' the animals' behaviour. Plainly, the proper way to avoid anthropomorphism is not to forswear the use of 'human' psychological descriptions altogether, but to exercise caution in their application, using them only where the evidence really warrants it.

Considered as sources of information about animal psychology, laboratory studies and field observations present complementary difficulties. A drawback of experimental work is that it provides no information about the lives of animals in their natural habitats; and so it gives us little knowledge about what the animals are like independent of human intervention. (How is the altruism that is displayed in the laboratory setting a part of the monkey's natural life?) For that information we must turn to the field studies. But those studies have their own problems, chief of which is the lack of experimental control: the field ethologists are, for the most part, passive observers of events and situations that they are powerless to manipulate; they are unable to bring about conditions that would confirm or disconfirm alternative hypotheses about the animals' behaviour. Limited to reporting whatever happens to pass before their eyes, their data are systematically incomplete and remain susceptible to rival interpretations. Thus when Jane Goodall reports a scene of apparent altruism, one is left wondering if a more modest interpretation might not explain the animal's behaviour just as well.

There is a way around these problems. The solution might be to combine what is learned in the laboratory with what is observed in the field. Experimental psychologists are able to exercise a degree of control that is impossible in the wild. Thus if monkeys are apparently altruistic, but one suspects that the explanation might be the influence of dominance–submissiveness relations, or fear of retaliation by larger animals, then the experiments can be repeated with these factors controlled. With this knowledge in hand, we can then turn to the field studies, and say with confidence that when the animals seem to display altruism in their native environments, it is reasonable to think that it really is altruism, and not a disguised version of something else. Alternative interpretations that are ruled out in the laboratory may be ruled out in Gombe and the Virungas as well.

Finally, if anthropomorphism is a sin, we should also be wary of the companion sin: the similarities between ourselves and other animals may too easily be *under*estimated. Consider, for example, the language used by the experimenters in reporting their findings about rhesus monkey altruism. The monkeys being tormented were called 'stimulus animals' and when they cried out in pain they were said to have 'vocalized'. The result of this was an 'increased noise level'—the language suggests that the animal pulling the chain was not even able to perceive the cries for what they were. The use of such language encourages us to picture the animals as considerably less complicated, and 'human', than they really are. When the great ethologist Konrad Lorenz was charged with anthropomorphizing the animals he studied, he replied, appropriately: 'You think I humanize the animals? . . . Believe me, I am not mistakenly assigning human propensities to animals; on the contrary, I am showing you what an enormous animal inheritance remains in man, to this day.' And this, of course, was Darwin's point as well.

RECAPITULATION

The main thesis of this book is that Darwinism leads inevitably to the abandonment of the idea of human dignity and the substitution of a different sort of ethic. In Chapter 2, I outlined the strategy that I would use in arguing for this claim. I would not argue that Darwinism entails the falsity of the doctrine of human dignity; rather, I would contend that Darwinism undermines human dignity by taking away its support.

The idea of human dignity is the moral doctrine which says that humans and other animals are in different moral categories; that the lives and interests of human beings are of supreme moral importance, while the lives and interests of other animals are relatively unimportant. That doctrine rests, traditiónally, on two related ideas about human nature: the idea that man is made God's image, and the idea that man is a uniquely rational being. In Chapter 3, I discussed the implications of Darwinism for religion and argued that if Darwinism is taken seriously the brand of theism that supports the image of God thesis is no longer a reasonable option. And now in Chapter 4, I have argued that Darwinism must also lead to the rejection of the idea that man is the only rational animal. We may now draw the conclusion that the traditional supports for the idea of human dignity are gone. They have not survived the colossal shift of perspective brought about by Darwin's theory.

It might be thought that this result need not be devastating for the idea of human dignity, because even if the traditional supports are gone, the idea might still be defended on some *other* grounds. Once again, though, an evolutionary perspective is bound to make one sceptical. The doctrine of human dignity says that humans merit a level of moral concern wholly different from that accorded to mere animals; for this to be true, there would have to be some big, morally significant difference between them. Therefore, any adequate defence of human dignity would require some conception of human beings as radically different from other animals. But that is precisely what evolutionary theory calls into question. It makes us suspicious of any doctrine that sees large gaps of any sort between humans and all other creatures. This being so, a Darwinian may conclude that a successful defence of human dignity is most unlikely.

5 Morality without the Idea that Humans are Special

I<small>F</small> the idea of human dignity is abandoned, what sort of moral view should be adopted in its place? This is not an easy question. Like Asa Gray, we are pulled in one way by the thought that we are kin to the animals, but we are pulled in a different direction by the conviction that, when all is said and done, we remain quite different from them. Anyone who attempts to formulate an adequate post-Darwinian ethic must feel this tension. The conflicting pulls are evident in the remarks of the distinguished evolutionary biologist Ernst Mayr:

The shockwaves of the 'dethroning' of man have not yet abated. Depriving man of his privileged position, necessitated by the theory of common descent, was the first Darwinian revolution. Like most revolutions, it went at first too far, as reflected in the claim made by some extremists that man is 'nothing but' an animal. This is, of course, not true. To be sure, man is, zoologically speaking, an animal. Yet, he is a unique animal, differing from all others in so many fundamental ways that a separate science for man is well-justified. When recognizing this, one must not forget in how many, often unsuspected, ways man reveals his ancestry. At the same time, man's uniqueness justifies, up to a point, a man-directed value system and man-centered ethics. In this sense a severely modified anthropocentrism continues to be legitimate.

We need a morality that will recognize both the similarities and the differences between humans and other animals, and that will be 'man-centered' only 'up to a point'. But what exactly is that point? How 'severely modified' must our anthropocentrism be? My aim in this chapter is to answer these questions. I will describe a view that I call moral individualism, and I will argue that it is the natural view for a Darwinian to adopt.

Moral individualism is a thesis about the justification of judgements concerning how individuals may be treated. The basic idea is that how an individual may be treated is to be determined, not by considering his group memberships, but by considering his own particular characteristics. If A is to be treated differently from B, the justification must be in terms of A's individual characteristics and B's individual characteristics. Treating them differently cannot be justified by pointing out that one or the other is a member of some preferred group, not even the 'group' of human beings.

This may seem excessively abstract, so let me give an example. Suppose we are considering whether we may use a chimpanzee in a medical experiment. In the course of the experiment the chimp will be infected with a disease and the progress of the disease will be observed; then he will be killed and his remains studied. In fact, such experiments have often been performed, and they are commonly considered to be morally acceptable. We may note, however, that the same experiment, performed on a human being, would not be considered acceptable. Appealing to the traditional doctrine of human dignity, we might explain this by saying that human life has an inherent worth that non-human life does not have. Moral individualism, on the other hand, would require a different approach. According to moral individualism, it is not good enough simply to observe that chimps are not members of the preferred group—that they are not human. Instead we would have to look at specific chimpanzees, and specific humans, and ask what characteristics they have that are relevant to the judgement that one, but not the other, may be used. We would have to ask what justifies using this particular chimp, and not that particular human, and the answer would have to be in terms of their individual characteristics.

This way of thinking goes naturally with an evolutionary perspective because an evolutionary perspective denies that humans are different in kind from other animals; and one cannot reasonably make distinctions in morals where none exist in fact. If Darwin is correct, there are no absolute differences between humans and the members of all other species—in fact, there are no absolute differences between the members of *any* species and all others. Rather than sharp breaks between species, we find instead a profusion of similarities and differences between particular animals, with the characteristics typical of one species shading over into the characteristics typical of another. As Darwin puts it, there are only differences of degree—a complex pattern of similarities and differences that reflect common ancestry, as well as chance variations among individuals within a single species. Therefore, the fundamental reality is best represented by saying that the earth is populated by individuals who resemble one another, and who differ from one another, in myriad ways, rather than by saying that the earth is populated by different *kinds* of beings. Moral individualism is a view that looks to individual similarities and differences for moral justification, whereas human dignity emphasized the now-discredited idea that humans are of a special kind.

Darwin emphasized the variety of similarities that exist between humans and other animals. In *The Expression of Emotion in Man and Animals*, he argued that animals experience anxiety, grief, dejection, despair, joy, love, tender feelings, devotion, ill-temper, sulkiness, determination, hatred, anger, disdain, contempt, disgust, guilt, pride, helplessness, patience, surprise, astonishment, fear, horror, shame, shyness, and modesty. As we have seen, much of Darwin's discussion of these themes might appear naïve to a reader today: he was ready to attribute fairly complex mental capacities to all sorts of animals on fairly slender grounds. Nevertheless, if he was right about even part of this, it would follow that, often, when we object to treating a human in a certain way, we would have similar grounds for objecting to the similar treatment of a non-human animal. If we think it is wrong to treat a human in a certain way, because the human has certain characteristics, *and a particular non-human animal also has those characteristics*, then consistency requires that we also object to treating the non-human in that way.

This line of reasoning provides some initial motivation for replacing the simple idea of human dignity with the more complex doctrine of moral individualism, according to which our treatment of individual creatures, human or non-human, should be adjusted to fit the actual characteristics of those creatures. A being's specific characteristics, and not simply its species membership, will then be seen as providing the basis for judgements about how it should be treated. This chapter is an elaboration of this idea, together with a fuller argument that it follows from an evolutionary perspective, and an outline of some of its practical consequences.

THE PRINCIPLE OF EQUALITY

We may begin by taking a brief look at the idea of human equality—an idea that lies near the heart of modern liberal thought, and that is espoused, in one form or another, by almost every Western thinker of the past three centuries. Despite its familiarity, it is not easy to say exactly what the ideal of equality is supposed to be. *Precisely* what is meant by saying that all persons are equal? Taken as a description of human beings, the claim that all are equal is plainly false. People differ in intelligence, beauty, talent, moral virtue, and physical strength—they differ in virtually every characteristic that might be thought important.

If the principle of equality is to be at all plausible, then, it cannot be interpreted as a factual statement describing how human beings are.

Instead, it must be understood as a principle governing how people are to be treated—as a moral rule saying, roughly, that people are to be *treated as* equals.

But there are problems with this, too. If people are not in fact equals, why should they be treated as such? Why should the smart and the stupid, the talented and the untalented, the virtuous and the vicious, be treated alike? This simple challenge has led some sceptical philosophers to abandon the idea of equality altogether, as a misguided ideal. The sceptics have a point, for surely we do not want all people to be treated in the same way. A doctor, for example, should not always prescribe the same treatment for every patient, regardless of what ails them. It would be a grisly joke always to prescribe penicillin on the 'egalitarian' grounds that this treats everyone alike. Again, should the admissions committee of a law school be required to admit (or reject) all applicants, because this treats them all as 'equals'? Obviously not: because all applicants are not equals, it makes no sense to treat them as though they were. At the same time, when people *are* equals—when there is no relevant difference between them—justice requires that they be treated similarly. This is just an application of the old Aristotelian point that like cases should be treated alike, and different cases differently.

We must formulate the principle of equality in such a way as to take this point into account. The basic idea that must be worked into the principle is that treating people differently is not objectionable if there is a relevant difference between them that justifies a difference in treatment. Thus, if one patient has an infection treatable by penicillin, while another patient does not, it is permissible to give one but not the other an injection of that drug. The difference between the patients justifies the difference in treatment. On the other hand, if there were no such difference between them—if both had exactly the same medical problem—there should be no difference in the treatment prescribed. Again, if one law-school applicant has a good college record, and has scored well on the qualifying examinations, while another applicant has a poor record, it is permissible to admit one but not the other. Again the difference between the people involved justifies the difference in treatment. Therefore, the principle of equality may be understood as saying that:

> Individuals are to be treated in the same way unless there is a relevant difference between them that justifies a difference in treatment.

But this only raises further questions. What, exactly, is a relevant difference? Suppose an employer will hire only whites, not blacks. He is basing his hiring policy on a difference between individuals—a difference in race—but is it a relevant difference that justifies the difference in treatment? If not, why not? Or suppose he pays women less than he pays men for the same work. Again, there is a difference—a difference in gender—but is it relevant, and if not, why not? Clearly, if it is permissible to cite any old difference between individuals as relevant, our principle is utterly empty, and a racist or sexist could happily accept it without altering their conduct one whit.

But not just any difference is relevant, and we do not have to accept race or gender as a legitimate basis for differential treatment simply because racists or sexists proclaim it as such. Whether a difference is relevant is a matter for rational assessment—we can ask why a difference between individuals justifies differential treatment, and if the difference is relevant, we may expect an answer.

To supplement the principle of equality, then, we need an explanation of what relevant differences *are*; we need a theory that specifies criteria for determining which differences are relevant and which are not. Such a theory would be the heart of any adequate theory of equality. I cannot develop a complete theory of relevant differences here—that would take us too far from the subject at hand, and it would involve controversies whose resolutions do not really matter for present purposes. But I do need to say something about what such a theory would be like.

In thinking about relevant differences, the first thing to notice is that whether a difference between individuals is relevant depends on the kind of treatment we have in mind. A difference between individuals that justifies *one* sort of difference in treatment might be completely irrelevant to justifying *another* difference in treatment. To continue with our earlier example, suppose the law-school admissions committee accepts one applicant but rejects another. Asked to justify this, they explain that the first applicant had excellent college grades and test scores, while the second applicant had a miserable record. Or suppose our doctor treats two patients differently: he gives one a shot of penicillin, and puts the other's arm in a plaster cast. Again, this can be justified by pointing to a relevant difference between them: the first patient had an infection while the second had a broken arm.

But now suppose we switch things around. Suppose the law school admissions committee is asked to justify admitting A while rejecting B, and replies that A had an infection but B had a broken arm. Or suppose the doctor is asked to justify giving A a shot of penicillin, while putting B's arm in a cast, and replies that A had better college grades and test scores. Both replies are, of course, silly, for it is clear that what is relevant in the one context is irrelevant in the other. The obvious point is that, before we can determine whether a difference between individuals is relevant to justifying a difference in treatment, we must know what sort of treatment is at issue. We might express this as a general principle:

> Whether a difference between individuals justifies a difference in treatment depends on the kind of treatment that is in question. A difference that justifies one kind of difference in treatment need not justify another.

Once this is made explicit, it seems obvious. But it has a corollary that is not so obvious: namely, that there is no *one* big difference between individuals that is relevant to justifying *all* differences in treatment. I say that this is not obvious because a lot of very smart people have overlooked it. Kant, for example, overlooked this point, and because of the oversight he made a crucial mistake that vitiated his whole discussion of the moral significance of species.

Kant, as we saw in Chapter 2, held that although we have duties involving animals, we can have no duties *to* animals—just as we can have duties involving trees, but not duties to trees. On Kant's view we may very well have a duty not to kill an animal—it may be someone's pet, for example, and the owner might be upset by its death. But the reason we should not hurt the animal has to do with the person's interests, not the animal's interests. On Kant's view, the animals' interests, considered by themselves, count for nothing at all.

What is the difference between humans and other animals that is supposed to justify this extraordinary difference in moral status? In one place, Kant says it is that humans are 'self-conscious', whereas other animals are not. (In another place, he implies it is because humans are autonomous agents, while other animals are not.) Here I am interested more in the form than in the content of Kant's proposal. He attempts to identify *one* difference between humans and non-humans that is relevant to justifying *all* differences in treatment. Consider the various ways in which non-humans are treated: we raise and eat them as food; we use them in laboratories, not only for medical and psychological experiments, but to test products such as soap and cosmetics; we dissect them in classrooms for educational purposes; we use their skins as clothing, rugs, and wall decorations; we make them objects of our amusement in zoos, circuses, and rodeos; we use them as work animals on farms; we keep them as pets; and we have a popular sport that consists of tracking them down and killing them for the pleasure of it. At the same time, we would think it deeply immoral if humans were treated in such ways. And Kant says, in effect, that this is all right, because mere animals are not self-conscious, and so we have no duties towards them. 7

But surely this cannot be right, for the characteristics that are relevant to justifying treatment vary with the different kinds of treatment. We admit humans, but not non-humans, to universities; and this is perfectly all right because the non-humans cannot read, write, or do mathematics. Here humans and animals are in different positions. But suppose we ask, not about admission to universities, but about torture: why is it wrong to cause an animal needless pain? The animal's inability to read, write, or do mathematics is irrelevant; what *is* relevant is its capacity for suffering. Here humans and non-humans are in the same boat. Both feel pain, and we have the same reason for objecting to torturing one as to torturing the other. *Very true – characteristic of pain both share.*

The strategy I am criticizing is common among contemporary moral philosophers, who assume that there must be some one big difference between us and the animals that puts us in different moral categories. Robert Nozick, author of the widely discussed book *Anarchy, State and Utopia*, holds that a being has rights only if it is a rational, free moral agent with 'the ability to regulate and guide its life in accordance with some overall conception it chooses to accept'. Lacking this ability, animals have no rights. Nozick is up to the old Kantian trick: the attempt to place a whole class of beings outside the sphere of morality, or at least the same part of morality we occupy, on the basis of their possession, or lack, of some one very general characteristic. But why should we believe that the same characteristics that are relevant to one form of treatment are also relevant to all the others? Surely the sensible approach is to take up the different forms of treatment, and the characteristics that make us eligible for them, one by one.

It might be objected that equality is an idea that applies only to humans, and that comparisons of humans with non-humans is therefore inappropriate. But the principle of equality, as I have formulated it here, seems to apply equally well to comparisons involving non-humans. Consider, for example, decisions regarding the treatment of chimpanzees versus the treatment of shrimp. Because a chimp is a curious, intelligent creature, it can easily suffer from boredom, and so some observers have criticized zoos for confining chimps in sterile, unstimulating environments. A chimp, they say, should not be placed in a bare cage with nothing to do but stare at the walls. Shrimp, however, are not curious and intelligent in the same way. Therefore a similar complaint could not be lodged about how they are treated. Because there is a relevant difference between them, it seems permissible to treat shrimp in ways that are objectionable where chimps are concerned.

Thus we find that when we compare humans with other humans, the principle of equality holds true, and when we compare animals with other animals, the principle also holds true. Why, then, shouldn't the principle be taken to apply with equal force to comparisons involving humans and other animals? Intuitively, the use of the principle in such contexts seems plausible. If it is permissible to treat a non-human animal in a certain way, while it is not permissible to treat a human in a similar way, surely there must be some difference between them that explains why. This is all that the principle of equality requires.

Let me add one other point about how relevant differences are recognized. Suppose the question is this. Rabbits have sometimes been used in procedures to test the safety of products such as shampoo. Before a new formula for shampoo is marketed for human use, we want to make sure the product is safe. People sometimes get shampoo in their eyes. Will this be harmful? A rabbit's eyes are similar to human eyes; so concentrated doses of the chemicals are applied to rabbits' eyes and the results observed. (This is called the Draize test.) If the results fall within certain limits, the product will be approved for human use. The procedure is obviously painful to the animals, who have to be restrained from rubbing the chemicals from their eyes. Now if *humans* were used in such tests, against their wills, there would undoubtedly be a great outcry. It is safe to say that there would be general agreement, even among those who find nothing wrong with treating rabbits in this way, that it is morally impermissible to do the same thing to humans. The question is, what is the difference between humans and rabbits that justifies this difference in treatment? *Pain + agitation*

Of course, there are many impressive differences between humans and rabbits. Humans, but not rabbits, can read; they can do mathematics; they can enjoy opera; they can drive automobiles; they can make movies. The list could go on and on. But are these differences relevant? Suppose they were cited as justification for permitting rabbits, but not humans, to be used in the Draize test. Would they be relevant differences? I suggest that this question can be answered as follows. First, forgetting rabbits for the moment, we ask why it would be objectionable to use humans in this way. The answer would be that the procedure is quite painful, and that people's eyes would be damaged beyond repair. This is bad for them because pain is bad, and because people need their eyes for all sorts of reasons—the loss of one's eyesight makes it more difficult to carry on one's life, regardless of the kind of life one has. Now we have identified the considerations that are relevant to justifying our judgement about humans. What is relevant is (*a*) the fact that humans are capable of suffering pain, and (*b*) the fact that humans need their eyesight in all sorts of ways for the conduct of their lives.

With this much in hand, we can then turn to the rabbits, and ask whether they are similar to humans in the relevant respects. Can they suffer pain? And do they need their eyesight to carry on their lives? If so, then we have the same reasons for opposing their use that we have for opposing the use of humans. And if someone objects that humans can do mathematics, or enjoy opera, but rabbits cannot, we can reply that even if these differences are relevant when other forms of treatment are at issue, they are irrelevant to the question about the Draize test.

The Draize test is only an example; more important, here at least, is the identification of a method for determining relevant differences. The procedure I have described may be stated in general terms: if it is thought permissible to treat A, but not B, in a certain way, we first ask *why* B may not be treated in that way. The reasons given will mention certain capacities of B. If A and B differ in that B lacks those capacities, then it is a relevant difference. But if A and B differ only in ways that do not figure in the explanation of why it is wrong to treat B in the specified manner, then the differences are irrelevant. This is far from a formal definition of relevant difference, but it does indicate something of how the concept works.

SPECIESISM

Recent writers on animal welfare have introduced the term 'speciesism' to refer to systematic discrimination against non-humans. (The term was coined by Richard Ryder, a British psychologist who quit experimenting on animals after he became convinced this was immoral, although Peter Singer's book *Animal Liberation* was responsible for popularizing the term.) Speciesism is said to be analogous to racism: it is the idea that the interests of the members of a particular species count for more than the interests of the members of other species, just as racism is the notion that the interests of the members of a particular race count for more. As Singer puts it:

The racist violates the principle of equality by giving greater weight to the interests of members of his own race when there is a clash between their interests and the interests of those of another race. The sexist violates the principle of equality by favoring the interests of his own sex. Similarly the speciesist allows the interests of his own species to override the greater interests of members of other species. The pattern is identical in each case.

The traditional doctrine of human dignity is speciesist to the core, for it implies that the interests of humans have priority over those of all other creatures. But let me try to be a little more precise about this.

Human speciesism can take two forms, one much more plausible than the other:

Radical speciesism: Even the relatively trivial interests of humans take priority over the vital interests of non-humans. Thus, if we have to choose between causing mild discomfort to a human, and causing excruciating pain to a non-human, we should prefer to cause pain to the non-human and spare the human.

always rules

This is the version of speciesism that Singer describes: one allows the interests of one's own species to override the *greater* interests of members of other species. We can, however, define a milder and more plausible version:

Mild speciesism: When the choice is between a relatively trivial human interest and a more substantial interest of a non-human, we may choose for the non-human. Thus it may be better to cause a little discomfort for a human than to cause agony for an animal. However, if the interests are comparable—say, if the choice is between causing the *same* amount of pain for a human or for a non-human—we should give preference to the human's welfare.

balance

Many defenders of traditional morality have embraced the radical form of speciesism. Aquinas and Kant, as we have seen, both held that the interests of non-humans count for nothing, and therefore may be outweighed by any human interest whatever. Indeed, on their view there is no point in doing any 'weighing' at all: the human always wins, no matter what. Descartes even denied that non-humans have any interests that *could* be weighed. Contemporary readers might find their views too extreme, and yet still find mild speciesism to be an attractive doctrine.

The principle of equality, on the other hand, involves the rejection of even mild speciesism: it implies that humans and non-humans are, in a sense, moral equals—that is, it implies that the interests of non-humans should receive the *same* consideration as the comparable interests of humans. I suspect that, viewed in this light, the principle of equality will seem implausible to many readers. The doctrine of human dignity, at least when it is interpreted as involving only mild speciesism, might appear to be a much more plausible view. Therefore, if I am to defend the principle of equality, I need to explain why even mild speciesism should be rejected.

Unqualified Speciesism

In addition to distinguishing between radical and mild speciesism, we may distinguish between qualified and unqualified versions of the *seperates into own categories.*

doctrine. The former distinction has to do with the extent of the view; the latter has to do with its logical basis.

Unqualified speciesism is the view that mere species alone is morally important. On this view, the bare fact that an individual is a member of a certain species, unsupplemented by any other consideration, is enough to make a difference in how that individual should be treated.

This is not a very plausible way of understanding the relation between species and morality, and generally it is not accepted even by those who defend traditional morality. To see why it is not plausible, consider the old science-fiction story 'The Teacher from Mars' by Eando Binder. The main character in that story is a Martian who has come to earth to teach in a school for boys. Because he is 'different'—seven feet tall, thin, with tentacles and leathery skin—he is taunted and abused by the students until he is almost driven out. Then, however, an act of heroism makes the boys realize they have been wrong, and the story ends happily with the ring-leader of the bullies vowing to mend his ways.

Written in 1941, the story is a not-so-thinly-disguised morality tale about racism. But the explicit point concerns species, not race. The teacher from Mars is portrayed as being, psychologically, exactly like a human: he is equally as intelligent, and equally as sensitive, with just the same cares and interests as anyone else. The only difference is that he has a different kind of body. And surely *that* does not justify treating him with less respect. Having appreciated this point, the reader is obviously expected to draw a similar conclusion about race: the fact that there are physical differences between whites and blacks—skin colour, for example—should make no moral difference either.

Although unqualified speciesism is implausible, as Binder's story shows, some philosophers have nevertheless defended it: they have argued that species alone *can* make a difference in our moral duties towards a being. Robert Nozick, for example, suggests that, in a satisfactory moral scheme,

perhaps it will turn out that the bare species characteristic of simply being human ... will command special respect only from other humans—this is an instance of the general principle that the members of any species may legitimately give their fellows more weight than they give members of other species (or at least more weight than a neutral view would grant them). Lions, too, if they were moral agents, could not then be criticized for putting other lions first.

Nozick illustrates the point with his own science-fiction example: 'denizens of Alpha Centauri' would be justified in giving greater weight to the interests of other such Alpha Centaurians than they give to our interests, he says, even if we were like them in all other relevant respects.

But this isn't at all obvious—in fact, it seems wrong on its face. If we substitute an Alpha Centaurian for a Martian in Binder's story, it makes no difference. Treating him less well merely because he is 'different' (in this case, a member of a different species) still seems like unjustified discrimination.

What of the 'general principle' Nozick suggests? It seems to be an expanded version of something that most people find plausible, namely, that one is justified in giving special weight to the interests of one's family or neighbours. If it is permissible to have special regard for family or neighbours, why not one's fellow species-members? The problem with this way of thinking is that there are lots of groups to which one naturally belongs, and these group-memberships are not always (if they are ever) morally significant. The progression from family to neighbour to species passes through other boundaries on the way—through the boundary of race, for example. Suppose it were suggested that we are justified in giving the interests of our own race greater weight than the interests of other races? Nozick's remarks might be adapted in defence of this suggestion:

> perhaps it will turn out that the bare racial characteristic of simply being white . . . will command special respect only from other whites—this is an instance of the general principle that the members of any race may legitimately give their fellows more weight than they give members of other races (or at least more weight than a neutral view would grant them). Blacks, too, could not then be criticized for putting other blacks first.

This would rightly be resisted, but the case for distinguishing by species alone is identical. As Binder's story suggests, unqualified speciesism and racism are twin doctrines.

Qualified Speciesism

But there is a more sophisticated view of the relation between morality and species, and it is this view that defenders of traditional morality have most often adopted. On this view, species alone is not regarded as morally significant. However, species-membership is correlated with *other* differences that *are* significant. The interests of humans are said to be more important, not simply because they are human, but because humans have morally relevant characteristics that other animals lack. This view might take several forms.

1. *The idea that humans are in a special moral category because they are rational, autonomous agents*. Humans, it might be said, are in a special moral category because they are rational, autonomous agents. Humans can guide their own conduct according to their own conceptions of what

ought to be done. (Since Kant, this has been the most popular way of describing the difference between humans and other animals, at least among philosophers.) It is this fact, rather than the 'mere' fact that they are human, that qualifies them for special consideration. This is why their interests are more important, morally speaking, than the interests of other species, although, it might be admitted, if the members of any other species were rational, autonomous agents, they would also go into the special moral category and would qualify for the favoured treatment. However, defenders of traditional morality insist that as a matter of fact no other species has this characteristic. So humans alone are entitled to full moral consideration.

Darwin, as we have seen, resisted the idea that humans have characteristics that are not shared by other animals. Instead he emphasized the continuities between species: if man is more rational than the apes, it is only a matter of degree, not of kind. But it may be of some interest to see what would follow *if* this were true. So let us set aside the Darwinian objection, and grant for the purpose of argument that humans are the only fully rational, autonomous agents. What would follow from this assumption?

Does the fact that someone is a rational autonomous agent make a difference in how he should be treated? Certainly it may. For such a being, the self-direction of his own life is a great good, valued not only for its instrumental worth but for its own sake. Thus paternalistic interference may be seen as an evil. To take a simple example: a woman might have a certain conception of how she wants to live her life. This conception might involve taking risks that we think are foolish. We might therefore try to change her mind; we might call attention to the risks and argue that they are not worth it. But suppose she will not heed our warnings: Are we then justified in forcibly preventing her from living her life as she chooses? It may be argued that we are not justified, for she is, after all, a rational, autonomous agent. It is different for someone who is *not* a fully rational being—a small child, for example. Then we feel justified in interfering with his conduct, to prevent him from harming himself. The fact that the child is not (yet, anyway) a fully rational agent justifies us in treating him differently from how we would treat someone who is a fully rational agent.

Of course, the same thing could be said to justify treating a human differently from a non-human. If we forcibly intervened to protect an animal from danger, but did not do the same for a human, we might justify this by pointing to the fact that the human is a rational autonomous agent, who knew what she was doing and who had the right to make her own choice, while this was not true of the animal.

Now notice two points about this reasoning. First, the fact that one individual is a rational autonomous agent, while another is not, sometimes justifies treating a human differently from a non-human, but it also justifies treating some humans differently from other humans. This consideration does not simply separate humans from animals; it separates humans from other humans as well. Thus, even if we grant (as a good Darwinian would not) that humans are the only rational, autonomous agents, we still have not identified a characteristic that separates all humans from all non-humans.

Secondly, and more important, once we understand *why* being a rational agent makes a difference in how one may be treated, in those cases in which it does make a difference, it becomes clear that possession of this quality is not always relevant. As we have already observed, whether a difference is relevant depends on the kind of treatment that is in question. When the issue is paternalistic interference, it is relevant to note whether the individual whose behaviour might be coerced is a rational agent. Suppose, however, that what is in question is not paternalistic interference, but putting chemicals in rabbits' eyes to test the safety of a new shampoo. To say that rabbits may be treated in this way, but humans may not, because human are rational agents, is comparable to saying that one law-school applicant may be accepted, and another rejected, because one has a broken arm while the other has an infection.

Therefore, the observation that humans are rational autonomous agents cannot justify the whole range of differences between our treatment of humans and our treatment of non-humans. It can justify some differences in treatment, but not others.

There is still another problem for this form of qualified speciesism. Some unfortunate humans—perhaps because they have suffered brain damage—are not rational agents. What are we to say about them? The natural conclusion, according to the doctrine we are considering, would be that their status is that of mere animals. And perhaps we should go on to conclude that they may be used as non-human animals are used— perhaps as laboratory subjects, or as food?

Of course, traditional moralists do not accept any such conclusion. The interests of humans are regarded as important no matter what their 'handicaps' might be. The traditional view is, apparently, that moral status is determined by what is normal for the species. Therefore, because rationality is the norm, even non-rational humans are to be treated with the respect due to the members of a rational species. Carl Cohen, a philosopher at the University of Michigan, apparently endorses this view in his defence of using animals, but not humans, in medical experiments. Cohen writes:

Persons who are unable, because of some disability, to perform the full moral functions natural to human beings are certainly not for that reason ejected from the moral community. *The issue is one of kind*. Humans are of such a kind that they may be the subject of experiments only with their voluntary consent. The choices they make freely must be respected. Animals are of such a kind that it is impossible for them to give or withhold voluntary consent or to make a moral choice. What humans retain when disabled, animals never had.

Let us pass over the obvious point that animals do seem to be able to withhold consent from participation in experiments—their frantic efforts to escape from the research setting, particularly when they are being caused acute discomfort, suggests that very strongly. But it is the more general theoretical point that we want to consider.

This idea—that how individuals should be treated is determined by what is normal for their species—has a certain appeal, because it does seem to express our moral intuitions about mentally deficient humans. 'We should not treat a person worse merely because he has been so unfortunate', we might say about someone who has suffered brain damage. But the idea will not bear close inspection. A simple thought-experiment will expose the problem. Suppose (what is probably impossible) that an unusually gifted chimpanzee learned to read and speak English. And suppose he eventually was able to converse about science, literature, and morals. Finally he expresses a desire to attend university classes. Now there might be various arguments about whether to permit this, but suppose someone argued as follows: 'Only humans should be allowed to attend these classes. Humans can read, talk, and understand science. Chimps cannot.' But this chimp *can* do those things. 'Yes, but *normal* chimps cannot, and that is what matters.' Following Cohen, it might be added that 'The issue is one of kind,' and not one of particular abilities accidental to particular individuals.

Is this a good argument? Regardless of what other arguments might be persuasive, this one is not. It assumes that we should determine how an individual is to be treated, not on the basis of *its* qualities, but on the basis of *other* individuals' qualities. The argument is that this chimp may be barred from doing something that requires reading, despite the fact that he can read, because other chimps cannot read. That seems not only unfair, but irrational.

2. *The idea that humans are in a special moral category because they can talk*. Traditionally, when Western thinkers characterized the differences between humans and other animals, the human capacity for language was among the first things mentioned. Descartes, as we have seen, thought that man's linguistic capacity was the clearest indication that he has a soul; and when Huxley was challenged by his working men to

explain why kinship with the apes did not destroy 'the nobility of manhood', he replied that 'man alone possesses the marvelous endowment of intelligible and rational speech.'

Is the fact that humans are masters of a syntactically complicated language, vastly superior to any communication-system possessed by non-humans, relevant to decisions about how they may be treated? In the preceding paragraphs I have already made some observations that bear on this. Clearly, it is sometimes relevant. It is relevant, for example, to the question of who will be admitted to universities. A knowledge of English is required to be a student in many universities, and humans, but not chimpanzees, meet this requirement. But not all humans qualify in this regard, and so it is reasonable to refuse admission to those humans. This means that it is the individual's particular linguistic capacity that is relevant to the admissions decision, and not the general capacities of 'mankind'. Moreover, there are many forms of treatment to which the question of linguistic ability is not relevant—torture, for example. (The reason why it is wrong to torture has nothing to do with the victim's ability to speak.) Therefore, the most that can be said about this 'marvelous endowment' is that *most* humans have it, and that it is relevant to *some* decisions about how they should be treated. This being so, it cannot be the justification of a principled policy of always giving priority to human interests.

It might be objected that this underrates the importance of language, because the implications of language are so diffuse. It isn't *simply* that knowing English enables one to read books, to ask and answer questions, to qualify for admission to universities, and so on. In addition to such discrete achievements, we have to consider the way that having a language enriches and extends all of one's other psychological capacities as well. A being with a language can have moral and religious beliefs that would otherwise be impossible; such a being's hopes, desires, and disappointments will be more complex; its activities will be more varied; its relationships with others will be characterized by greater emotional depth; and on and on. In short, its whole life will be richer and more complex. The lives of creatures who lack such a language will be correspondingly simpler. In light of this, it will be argued, isn't it reasonable to think that human language makes human life morally special?

There is obviously something to this. I think it is true that possession of a human language enriches almost all of one's psychological capacities; that this has consequences that ramify throughout one's life; and that this is a fact that our moral outlook should accommodate. But it is not obvious exactly how this fact should figure into our moral view. What, exactly, is its significance? It does not seem right to say that, because of this, human interests should always have priority over the interests of non-humans, for there may still be cases in which even the enriched capacities of humans are irrelevant to a particular type of treatment. I want to make a different suggestion about its significance.

Suppose the type of treatment in question is killing: say, we have to choose between causing the death of a human, and causing the death of a non-human animal. On what grounds may this choice be made? Although killing is a specific type of treatment, its implications are especially broad: one's death puts an end to all one's activities, projects, plans, hopes, and relationships. In short, it puts an end to one's whole life. Therefore, in making this decision it seems plausible to invoke a broadly inclusive criterion: we may say that the *kind of life* that will be destroyed is relevant to deciding which life is to be preferred. And in assessing the kind of life involved, we may refer, not just to particular facts about the creatures, but to summary judgements about what all the particular facts add up to. Humans, partly because of their linguistic capacities, have lives that are richer and more complex than the lives of other animals. For this reason, one may reasonably conclude that killing a human is worse than killing a non-human.

If this account is correct, it would also explain why it is worse to kill some non-humans than others. Suppose one had to choose between killing a rhesus monkey and swatting a fly. If we compare the two, we find that the life of the monkey is far richer and more complex than that of the fly, because the monkey's psychological capacities are so much greater. The communicative abilities of the monkey, we may note, also make an important difference here. Because the monkey is able to communicate with others of its own kind—even though its communicative skills are inferior to those of humans—its relations with its peers are more complex than they would otherwise be. (This is a clear illustration of Darwin's thesis that the differences between humans and non-humans are matters of degree, not kind.) In light of all this, we may conclude that it is better to swat the fly. This result is intuitively correct, and it lends additional plausibility to the general idea that, where killing is concerned, it is the richness and complexity of the life that is relevant to judgements about the wrongfulness of its destruction.

This is compatible with moral individualism only if we add a certain qualification, namely, that it is the richness and complexity of the *individual* life that is morally significant. Some humans, unfortunately, are not capable of having the kind of rich life that we are discussing. An infant with severe brain damage, even if it survives for many years, may never learn to speak, and its mental powers may never rise above a primitive level. In fact, its psychological capacities may be markedly inferior to those of a typical rhesus monkey. In that case, moral individualism would see no reason to prefer its life over the monkey's. This will strike many people as implausible. Certainly, the traditional doctrine of human dignity would yield a different result. Nevertheless, I think that moral individualism is correct on this point, and I will have more to say about this below.

3. *The idea that humans are in a special moral category because they alone are able to participate in the agreements on which morality depends.* A different sort of argument turns on a certain conception of the nature of morality and the sources of moral obligation. This argument is connected with the intuitively appealing idea that human beings are members of a common moral community; that morality grows out of their living together in societies and co-operating to provide for their common welfare. This creates bonds between them in which non-humans have no part. Thus, humans have obligations to one another that are importantly different from any obligation they might have to mere animals.

Spelled out in greater detail, this argument depends crucially on the notion of *reciprocity*. It is plausible to think that moral requirements can exist only where certain conditions of reciprocity are satisfied. The basic idea is that a person is obligated to respect the interests of others, and acknowledge that they have claims against him, only if the others are willing to respect his interests and acknowledge his claims. This may be thought of as a matter of fairness: if we are to accept inconvenient restrictions on our conduct, in the interests of benefiting or at least not harming others, then it is only fair that the others should accept similar restrictions on their conduct for the sake of our interests.

The requirement of reciprocity is central to contract theories of ethics. Such a theory conceives of moral rules as rules which rational, self-interested people will agree to obey on condition that others will obey them as well. Each person can be motivated to accept such an arrangement by considering the benefits he will gain if others abide by the rules; and his own compliance with the rules is the fair price he pays to secure the compliance of others. That is the point of the 'contract' which creates the moral community.

This conception helps us to understand, easily and naturally, why we have the particular moral rules we do. Why do we have a rule against killing? Because each of us has something to gain from it. It is to our advantage that others accept such a rule; for then we will be safe. Our own agreement not to harm others is the fair price we pay to secure their agreement not to harm us. Thus the rule is established. The same could be said for the rule requiring us to keep our promises, to tell the truth, and so on.

It is a natural part of such theories that non-human animals are not covered by the same moral rules that govern the treatment of humans, for the animals cannot participate in the mutual agreement on which the whole set-up depends. Thomas Hobbes, the first great social contract theorist, was well aware of this: 'To make covenants with brute beasts', he said, 'is impossible.' This implication is also made explicit in the most outstanding recent contribution to contract theory, John Rawls's *A Theory of Justice*. Rawls identifies the principles of justice as those which would be accepted by rational, self-interested people in what he calls 'the original position'; that is, a position of ignorance with respect to particular facts about oneself and one's position in society. The question then arises as to what sorts of beings are owed the guarantees of justice, and Rawls's answer is:

We use the characterization of the persons in the original position to single out the kinds of beings to whom the principles chosen apply. After all, the parties are thought of as adopting these criteria to regulate their common institutions and their conduct toward one another; and the description of their nature enters into the reasoning by which these principles are selected. Thus equal justice is owed to those who have the capacity to take part in and to act in accordance with the public understanding of the initial situation.

This, he says, explains why non-human animals do not have the 'equal basic rights' possessed by humans; 'they have some protection certainly but their status is not that of human beings'. And of course this result is not surprising: for if rights are determined by agreements of mutual interest, and animals are not able to participate in the agreements, then how could *their* interests possibly give rise to 'equal basic rights'?

The requirement of reciprocity may seem plausible, and I think that it does contain the germ of a plausible idea—I will say more about this in a moment—but nevertheless there are good reasons to reject it. To see why, we need to distinguish the conditions necessary for *having* a moral obligation from the conditions necessary for being the *beneficiary* of a moral obligation.

For example: normal adult humans have the obligation not to torture one another. What characteristics make it possible for a person to have this obligation? For one thing, he must be able to understand what torture is, and he must be capable of recognizing that it is wrong. (Linguistic capacity might be relevant here; without language one may not be able to formulate the belief that torture is wrong.) When someone—a severely retarded person, perhaps—lacks such capacities, we do not think he has such obligations and we do not hold him responsible for what he does. On the other hand, it is a very different question what characteristics qualify someone to be the beneficiary of the obligation. It is wrong torture someone—someone is the beneficiary of our obligation not to torture—not because of his capacity for understanding what torture is, or for recognizing that it is morally wrong, but simply because of his capacity for experiencing pain. Thus a person may lack the characteristics necessary for *having* a certain obligation, and yet may still possess the characteristics necessary to qualify him as the *beneficiary* of that obligation. If there is any doubt, consider the position of severely retarded persons. A severely retarded person may not be able to understand what torture is, or see it as wrong, and yet still be able to suffer pain. So we who are not retarded have an obligation not to torture him, even though he cannot have a similar obligation not to torture us.

The requirement of reciprocity says that a person is morally obligated to accept restrictions on his conduct, in the interests of not harming others, only if the others reciprocate. The example of the retarded person shows this to be false. He is not capable of restricting his conduct in this way; nevertheless we have an obligation to restrict ours. We are in the same position with respect to non-human animals: like the retarded person, they may lack the characteristics necessary for having obligations; but they may nevertheless qualify as beneficiaries of our obligations. The fact that they cannot reciprocate, then, need not affect our basic obligations to them.

I said that the requirement of reciprocity, although unacceptable, does contain the germ of a plausible idea. What I have in mind is the idea that if a person *is* capable of acting considerately of our interests, and *refuses* to do so, then we are released from any similar obligations we might have had to him. This may very well be right. But whether or not this point is accepted makes no difference to our duties to non-human animals, since they lack the capacity to 'refuse' to recognize obligations to us, just as they are not able to accept such obligations.

There is one other way that considerations of reciprocity might enter into one's decisions about what to do. Suppose that, at some time in the past, a particular person has done you a good turn. You might consider yourself to be indebted to that person, so that if you have the chance to be helpful to him in the future, you have a special obligation to do so. Thus, if you have to choose between helping him, and helping someone to whom you owe no such debt, you may legitimately choose in favour of your benefactor. (This may be a 'relevant difference' between them that satisfies the demands of the principle of equality.) There is no objection to this, but at the same time it provides no particular grounds for distinguishing between one's general obligations to humans and one's general obligations to other animals. Rather it is a principle that comes into play most often in distinguishing between what one owes to different humans; and moreover, one *could* sometimes have a special obligation of this sort to a non-human. After all, non-humans have on occasion performed valuable services for humans, and it would be ungrateful to think that they could never deserve any credit for this.

4. *The idea that humans are in a special moral category because they are more sensitive to harm than other creatures.* Finally, we need to consider briefly an argument that is not very impressive, but that one sometimes hears. I have said (several times in fact) that even though there are frequently important differences between humans and other animals, there may be no difference in their capacities for experiencing pain. Humans and non-humans both suffer; and so, regardless of what other differences may exist, we have the same basic reason for objecting to tormenting an animal that we would have for objecting to tormenting a human. At this fundamental level, humans and non-humans are surely equal.

But it might be objected that this is not so, that humans suffer more than other animals when they are caused distress. Because of their capacity of foresight, humans can anticipate painful experiences and dread them in advance. This dread can have a vivid, detailed quality: one knows, not simply that one will suffer, but that one will suffer in particular ways; and afterwards, the memory may remain to haunt one indefinitely. Animals with less extensive cognitive abilities will suffer fewer of these ancillary effects. This applies not only to physical suffering, but to psychological suffering as well. A human mother, forcibly separated from her child, may grieve the loss for the rest of her life. A female rhesus monkey whose baby is taken away might also be traumatized, but she will soon get over it. One cannot, therefore, equate the mistreatment of a human with the mistreatment of an animal, even when they appear on the surface to be similar.

There is obviously something to this; but it is important to understand what follows from it and what does not. Nothing in this line of reasoning invalidates the fundamental idea that the interests of non-humans should receive the same consideration as the *comparable* interests of humans. All that follows is that we must be careful in assessing when their interests are really comparable. The situation may be represented schematically as follows. Suppose we must choose between causing x units of pain for a human or a non-human. Because of the human's superior cognitive abilities, the after-effects for him will include y additional units of suffering; thus the human's total misfortune will be $x + y$, while the non-human's total will be only x. Thus the human has more at stake, and the principle of equality would recommend favouring him.

On the other hand, suppose we alter the example to make the non-human's initial pain somewhat more intense, so that it equals $x + y$. Then the total amounts of suffering would once again be comparable, and the point about the human's greater sensitivities would not provide any justification for preferential treatment.

Qualified speciesism is the view that the interests of humans are morally more important, not simply because they are human, but because humans have morally significant characteristics that other animals lack. But what are those characteristics? We have now considered several possibilities: that humans can speak; that they are rational agents; that they are moral agents, capable of having obligations; that they are capable of entering into agreements of mutual benefit with other humans, and performing services for them; and that they are more vulnerable to harm. In examining each of these, we have found no reason to abandon the approach suggested by the principle of equality: where relevant differences between individuals exist, they may be treated differently; otherwise, the comparable interests of individuals, whether human or non-human, should be given comparable weight. We have found no reason to support a policy of distinguishing, in principle, between the kind of consideration that should be accorded to humans and that which should be accorded to other animals.

Where does this leave the relation between species and morality? The picture that emerges is more complex, but also more true to the facts, than traditional morality. The fact is that human beings are not simply 'different' from other animals. In reality, there is a complex pattern of similarities and differences. The matching moral idea is that in so far as a human and a member of another species are similar, they should be treated similarly, while to the extent that they are different they should be treated differently. This will allow the human to assert a right to better treatment whenever there is some difference between him and the other animal that justifies treating him better. But it will not permit him to claim greater rights simply because he is human, or because humans in general have some quality that he lacks, or because he has some characteristic that is irrelevant to the particular type of treatment in question.

THE CONNECTION BETWEEN DARWINISM AND MORAL INDIVIDUALISM

There is a striking parallel between this way of understanding morality and Darwin's view about the nature of species. Before Darwin, when species were thought to be immutable, naturalists believed that membership in a species was determined by whether the organism possessed the qualities that defined the essence of the species. For the pre-Darwinian naturalist, variations were of little interest, except as curiosities. It was, after all, the 'standard' specimen that exemplified the external essence of the species, which the naturalist was trying to learn about. This essence was something real and determinate, fixed by nature itself, and the systems of classification devised by biologists were viewed as accurate or inaccurate depending on how well they corresponded to the fixed order of nature.

Evolutionary biology implies a very different view. Darwin argued that there are no fixed essences; there is only a multitude of organisms that resemble one another in some ways but differ in others. (Moreover, variations are no longer to be regarded as mere curiosities; on the contrary, they are the very stuff of nature—they are what make natural selection possible.) How those individuals are grouped—into species, varieties, and so on—is more or less arbitrary. In *The Origin of Species* Darwin declared:

I look at the term species, as one arbitrarily given for the sake of convenience to a set of individuals closely resembling each other, and that it does not essentially differ from the term variety, which is given to less distinct and more fluctuating forms. The term variety again, in comparison with mere individual differences, is also applied arbitrarily, and for mere convenience sake.

Thus Darwinian biology substitutes individual organisms, with their profusion of similarities and differences, for the old idea of determinate species; while moral individualism substitutes the view that our treatment of those organisms must be sensitive to those similarities and differences, for the old view that what matters is the species to which the organism belongs.

There is, then, a sense in which moral individualism 'fits' well with evolutionary theory. We may wonder, however, exactly what this 'fit' consists in. The relation between the two is surely weaker than logical entailment. Darwinian theory does not strictly entail the truth of moral individualism. One could be a Darwinian, and reject moral individualism, without falling into self-contradiction. On the other hand, the connection seems stronger than mere consistency, which would require only that there be no formal contradiction in accepting both. What, then, is the relation between them?

It will be helpful to look again at the connection in traditional morality between the image of God thesis, the rationality thesis, and the idea of human dignity. Traditional morality proceeded from the assumption that human beings are ineluctably 'different' from other animals; the difference was usually explained by saying that humans alone were made in the image of God, or that humans alone are rational, or both. It was assumed that this conferred upon human life, and human interests, a special importance which raised their status far above that of mere animals. But why do these 'facts' (that men are made in the image of God, or that they are uniquely rational) confer a special moral status upon humans? What is the connection between them?

In Chapter 2, I emphasized that the doctrine of human dignity is not an isolated moral idea; it 'fits' the pre-Darwinian understanding of humans as being different in kind from other creatures. More specifically, I suggested that the image of God thesis and the rationality thesis functioned as *reasons* in support of the doctrine of human dignity. We are now in a position to add an important detail to this account. The additional detail has to do with the principle of equality.

The principle of equality, as I have elaborated it, is not a new idea. The realization that the principle requires the abandonment of slavery, racism, and sexism is relatively recent; but the principle itself is a principle of rationality that has been operating for as long as people have been capable of being reasonable. (Aristotle knew that like cases should be treated alike, and different cases should be treated differently; so when he defended slavery he felt it necessary to explain why slaves are 'different'.) Therefore, if the doctrine of human dignity was to be maintained, it was necessary to identify the difference between humans and other animals that justified the difference in moral status. That is where the image of God thesis and the rationality thesis came in. They supplied the needed relevant differences.

Therefore, the 'fit' between the image of God thesis, the rationality thesis, and the doctrine of human dignity can be explained like this: they go together naturally because they form parts of a chain of reasoning, in which the principle of equality plays a key role. The principle of equality mediates the fit between them.

> Individuals are to be treated in the same way, unless there is a difference between them that justifies a difference in treatment.
>
> Humans and other animals are radically different in kind: humans alone are made in the image of God, and humans alone are rational.
>
> Being made in the image of God, and being rational, are characteristics that justify special treatment: creatures that have these characteristics are to be treated with special concern and respect.
>
> Therefore, humans and other animals are to be treated in radically different ways. The interests of humans are to be accorded a level

of concern far above the concern shown for the interests of any other creatures.

The 'fit' between an evolutionary perspective and moral individualism can be explained in a similar manner. We learn from Darwin that, contrary to what was previously believed, humans and other animals are not radically different in kind; and with this new understanding we are compelled to reason differently:

> Individuals are to be treated in the same way, unless there is a difference between them that justifies a difference in treatment.

> Humans and other animals are not radically different in kind: they are similar in some ways, and different in others, and these differences are often merely matters of degree. If humans are rational, so are other animals, although perhaps to a different degree. The same goes for other important human capacities.

> Therefore, when humans have characteristics that justify treating them in certain ways, it may be that other animals also have those characteristics.

> Therefore, our treatment of humans and other animals should be sensitive to the pattern of similarities and differences that exist between them. When there is a difference that justifies treating them differently, we may; but when there is no such difference, we may not.

It is the principle of equality that mediates the 'fit' between moral individualism and an evolutionary outlook. Moral individualism is, therefore, nothing but the consistent application of the principle of equality to decisions about what should be done, in light of what Darwinism has taught us about our nature and about our relation to the other creatures that inhabit the earth.

THE VALUE OF HUMAN LIFE

The big issue in all this is the value of human life. Darwin's early readers—his friends as well as his enemies—worried that, if they were to abandon the traditional conception of humans as exalted beings, they could no longer justify the traditional belief in the value of human life. They were right to see this as a serious problem. The difficulty is that Darwinism leaves us with fewer resources from which to construct an account of the value of life. Traditional theorists could invoke mankind's divine origins and special place in God's plan, as well as the idea that human nature is radically different from animal nature. Using these notions, they could devise a robust account of the sanctity of human life and its consequent inviolability. A Darwinian must make do with skimpier materials. With the old resources no longer available, one might well wonder whether we are left with enough to construct a viable theory.

What would a post-Darwinian theory of the value of human life be like? On what could it be based? Such a theory might begin by emphasizing that the value of a life is, first and foremost, the value that it has *for the person who is the subject of that life*. Our lives are valuable, not to God or to nature or to the universe, but to us. This thought, familiar from the writings of religious sceptics, is not so simple as it first appears. It may easily be misunderstood, and we need to be as clear as possible about its meaning.

When we say that something is valuable 'to someone', we might mean three things. First, we might mean that someone *believes* the thing is valuable; or secondly, we might mean that someone consciously *cares* about it. These interpretations suggest that a subjectivist theory of value is being assumed. But there is a third, more straightforward under-standing of what is meant, that does not involve any sort of subjectivist understanding of value. In saying that something is valuable for some-one, we might simply mean that *this person would be worse off without it*. It is important to notice that this third sense is independent of the first two. The loss of something might in fact be harmful to someone even though they were ignorant of this fact and consequently did not care about it.

Consider the way in which my eyes are valuable 'to me'. If I were struck blind, this would be a bad thing, not for the universe, but *for me*. It would be needlessly obscure to try to account for my misfortune by appealing to some abstract standard which attributes evil to impersonally considered states of nature. The simple, obvious explana-tion is best: the loss of my eyesight would be a bad thing for me because I would be harmed by it. Of course, this might also be a bad thing for those around me, who would be adversely affected by my incapacity. But we may take this as a secondary matter; the primary evil consists in the harm that *I* would suffer.

Similarly, when we say that a woman's life is valuable 'to her', it is this third sense that is intended: her life is valuable to her because *she is the one who would be harmed by its loss*. If she is killed, she has lost something— her life—that is of supreme importance, not to nature or to the universe, but to her. Her death might also be a bad thing for her friends and family, but once again this is secondary. The primarily evil consists in the loss that *she* suffers. If this is correct, then we are in a position to account for the value of life without having to invoke a divine order or an overly inflated conception of what human beings are like. It is enough to attend to the kind of value that our lives have for us.

Why, exactly, is the loss of life harmful? To understand this we need to distinguish two notions that are often conflated: we need to separate *being alive* from *having a life*. The former is a notion of biology: to be alive is to be a functioning biological organism; it is the opposite of being dead, or of being the kind of thing that is neither alive nor dead, such as a rock. The latter is a notion not of biology but of biography. Consider, for example, the life of John Dalton Hooker. Hooker was born in 1817, the son of a distinguished botanist. After completing his medical studies, he sailed to Antarctica on HMS *Erebus*. He was introduced to Darwin in 1839 and became his close friend. The first of his two wives was Frances Henslow, the professor's daughter. For twenty years he was director of the Royal Botanic Gardens, having succeeded his own father in that post. For his work as a naturalist he was awarded the Royal Medal of the Royal Society, the Copley Medal, and in 1892 the Darwin Medal. He was a pallbearer at Darwin's funeral, and finally retired to Berkshire where he died in 1911. These are some of the facts of Hooker's life. They are not biological facts, although some of them might involve biology. Primarily they are facts about his history, character, actions, interests, and relationships.

Once this distinction is made clear, we can see that there is a deep ambiguity in the notion of the value of life. Which is important to us—life in the biological sense, or life in the biographical sense? Plainly, the latter seems more important. Our lives are the sum of all we hold dear: our projects, our activities, our loves and friendships, and all the rest. Being alive, by contrast, is valuable to us only in so far as it enables us to carry on our lives. This is most evident when we consider the extreme case in which a person, while still alive, has lost the capacity for having a life, such as a person in irreversible coma. Being alive, sadly, does such a person no good at all. The value of being alive may therefore be understood as instrumental; being alive is important to an individual because it enables him or her to have a life.

This suggests that the moral rule against killing may be understood as a derived rule. To kill someone is to destroy a biological life; this is objectionable because, without biological life, there can be no biographical life. (Being blind makes one's life more difficult; being dead makes it impossible.) The point of the rule against killing is, therefore, to protect individuals against the loss of their lives, and not merely against the cessation of their being alive.

But an adequate theory should do more than specify the content of the rule against killing: it should also explain why such a rule should be accepted. After all, the acceptance of any moral rule involves putting up with inconvenient constraints on one's behaviour: to accept the rule against killing means that one may not kill another person, even if it would be to one's own advantage to do so. Why should such a rule be accepted? Why, it might be asked, should anyone take the fact that other people value their lives, in the sense outlined here, as a compelling reason for accepting inconvenient restrictions on what may be done? Part of the answer is provided by the Darwinian account of the 'social instincts'. We are social animals, and the capacity for caring about the welfare of others is part of our nature. But at best this only explains why we *do* accept the rule; it does not explain why we *should* accept it. The rest of the explanation is provided by the principle of equality. Each of us—or at least, each normal person among us—is the subject of a life, and each of us would be harmed in the same way by its loss. In this respect we are all in the same boat; there is no relevant difference between us. Therefore, if we think that others should not kill us, we have to acknowledge that we should not kill them.

Finally, an adequate theory should offer guidance in making actual moral decisions. At some point, theory and practice should come together. What are the practical consequences of the more modest view of the value of life that I have outlined? As one might suspect, the practical consequences of the more modest view are markedly different from the prescriptions of traditional morality. This is evident when we consider such matters as suicide. In dealing with this issue, and others like it, we will have to avoid relying on arguments that smuggle in pre-Darwinian conceptions of human nature. There is a whole family of arguments that appeal, directly or indirectly, to the idea that human beings are 'a great work, worthy the interposition of a deity'. These arguments might refer to the dignity of man, to the sacredness of human life, or to some other noble-sounding principle. A 'more humble' view of ourselves requires that we set these arguments aside. Whenever they have a legitimate point, the point will have to be restated in terms of the specific human characteristics that are relevant to justifying the particular moral judgement. If the arguments cannot be recast in these more modest terms, then the slogans—however they are phrased—must be rejected as mere puffery.

In his discussion of suicide Kant appealed to arguments that look quite suspicious when viewed in this light. Kant, as we saw in Chapter 2, held that 'suicide is in no circumstances permissible'. He wrote:

Its advocates argue thus. So long as he does not violate the proprietary rights of others, man is a free agent. With regard to his body there are various things he can properly do; he can have a boil lanced or a limb amputated, and disregard a scar; he is, in fact, free to do whatever he may consider useful and advisable. If he then comes to the conclusion that the most useful and advisable thing that he can do is to put an end to his life, why should he not be entitled to do so? Why not, if he sees that he can no longer go on living and that he will be ridding himself of misfortune, torment and disgrace?

To make the argument less abstract, let us consider an example. Richard Schlatter was a Rhodes scholar who rose to prominence as a teacher and administrator at Rutgers University. From 1962 to 1972 he was provost of Rutgers and a good friend of the president, Mason Gross. Gross died in 1972 after a protracted illness filled with suffering that, according to newspaper accounts, was prolonged 'past any sensible point'. Dr Schlatter was deeply affected by his friend's death, and began to vow that, if he were ever in such a position, he would end his life. Then in 1987, at age 75, he found himself in exactly this position after he was diagnosed as having terminal cancer of the spine. In considerable pain, and concluding that his condition was hopeless, Dr Schlatter shot himself to death. The 'advocates', as Kant calls them, might argue that Dr Schlatter's decision was rational and within his rights. His biographical life was, to all good purposes, over, and his biological life was no longer of any use to him—indeed, his own opinion was that it had become an intolerable burden.

But Kant would have none of this. His reason was that suicide is a violation of the nobility of human nature. 'Man is not a thing,' he said, 'not a beast. If he disposes over himself, he treats his value as that of a beast.'

Suicide is not abominable and inadmissible because life should be prized; were it so, we could each have our own opinion of how highly we should prize it, and the rule of prudence would often indicate suicide as the best means. But the rule of morality does not admit of it under any condition because it degrades human nature below the level of animal nature and so destroys it.

How, exactly, does the suicide 'treat his value as that of a beast'? This is not at all clear, but Kant seems to think that the very existence of rational beings is a good thing, while the existence of non-rational creatures has no comparable value. (He did not, of course, believe that non-human animals are rational beings.) We might therefore interpret his argument as saying that one has a duty not to decrease the number of rational beings in the universe, and that this duty overrides any considerations having to do with what is good for oneself.

Interpreted in this way, Kant's argument appeals to an understanding of the value of human life according to which the existence of humans is somehow good for the universe—it is better, in some abstract way, for there to be people than for there not to be people, and the more people the better. This is in sharp contrast to our view that the value of a human life is primarily a value that it has for the person whose life it is. Yet how could the Kantian view be sustained? It is hard to see how, without the resources of which Darwin deprives us, one could justify such a view.

Kant's comparison of how we may treat humans with how we may treat beasts also recalls another old argument. Those who think that suicide is sometimes permissible have often urged that there is no difference between shooting a horse to put it out of misery and killing a man for the same reason. If we think that one is acceptable for humane reasons, then why not the other? Part of Kant's point seems to be that we cannot apply the same reasoning to humans as to horses because humans and horses are different kinds of beings. Treating a human as one would treat a horse is not permissible because—well, because a man is a man and a horse is a horse.

There are two ways in which this can be understood. The less charitable interpretation would be that Kant is appealing, once again, to human *hubris*—to our sense that human beings have too grand a natural status to be treated as one would treat mere animals. A more charitable understanding, however, would interpret his claim that human nature is different from animal nature, in Darwinian terms, as shorthand for the observation that humans typically have sets of characteristics that other animals lack, and vice versa. Then, remembering the principle of equality, we could ask whether the particular characteristics that justify killing horses are shared by humans. Can this version of Kant's argument be recast in more modest terms? If so, then it would be acceptable to moral individualism as well as to traditional thinking.

The question is whether there are differences between horses and humans that would justify taking such different attitudes towards them. We may suppose that the horse has suffered a fatal injury that cannot be remedied, and that, like Dr Schlatter, the human has a painful terminal disease. The similarities are obvious: both are dying; neither has the prospect of a future life; and both are suffering. Even so, of course, there may be differences: the human, but not the horse, might have the capacity of understanding what is happening to him and might express a preference about it. If the man does not want a quicker death, then we have a powerful reason against treating him as we would treat the horse. Moreover, the man might choose to kill himself, while the horse lacks this ability. Still, if the man expresses a preference for a quick and painless death, and if he seeks to bring this about by committing suicide, this does not weaken the case for treating them alike—on the contrary, it strengthens it.

Therefore, when we consider the specific characteristics of human beings, we find nothing that can form the basis of an absolute prohibition against self-destruction. Kant's appeal to 'degrading human nature' seems, therefore, to be no more than puffery—when the emotional rhetoric is stripped away, nothing is left.

An ethic that appeals only to what is good for people will not endorse an absolute prohibition of suicide. This does not mean, however, that suicide will be taken lightly. In practice, rational suicides are rare—so rare that wise people are reluctant ever to recommend it, for fear of causing tragedy. Most actual suicides are tragic because they occur when people are despondent and are taking an unreasonably pessimistic view of their prospects. It should be emphasized that, even on our more modest view of the value of life, someone's life may be valuable to them even if they do not believe it is valuable and even if they do not consciously care about it. In a typical case, a man may be depressed and despair that his life is no longer worth living. The depression may be only temporary. For the time being, however, he has ceased to believe his life is worthwhile, and he does not care whether he lives or dies. But even while he is in this state, we can sensibly say that his life is valuable, and not just that it is valuable in the abstract, but that it is valuable *to him*, for it is a fact, unaffected by his present attitude, that the loss of his life would do him great harm. We cannot assume, though, that *every* case is like this. Sadly, there may be some cases in which the gloomy estimate is correct; and in those cases we may have to concede, however reluctantly, that suicide is rational.

The question of suicide is closely related to the question of euthanasia, and our theory of the value of life will treat them similarly. Suppose that, like Patricia Rosier, a dying person has been reduced to such an intolerable level of existence that she prefers to die quickly rather than to linger on for a few more miserable days. Mrs Rosier was an avid tennis player, jogger, and butterfly collector who contracted cancer and developed four malignant tumours in her brain. She appeared on television in her home town of Ft Myers, Florida, several times in 1985, discussing her cancer and offering advice to help others deal with terminal illness. Finally, as her condition was worsening, she asked her husband, a physician, to help her die painlessly. He gave her a lethal dose of morphine, reported what he had done, and was then charged with first-degree murder.

Setting aside questions of legality, is mercy killing, in such circumstances, morally wrong? It has been opposed on various grounds, including the fear that acceptance of euthanasia would lead inevitably to a diminished respect for life and so to further killings that no one would want. This objection, however, leaves the permissibility of mercy killing a contingent matter, to be decided on the basis on one's estimate of causes and effects. The opponents of euthanasia have rarely been willing to leave it at that. The deeper reason for their opposition, which suggests a more unconditional prohibition, is reverence for human life.

'Human life', declared one recent writer on the subject, 'is of infinite value.' Reading such words, one might assume that they are not meant to be taken literally. *Infinite* value seems too much, even for such a precious thing as human life. But this particular writer—Dr Moshe Tendler, a professor of Talmudic law—means precisely what he says. Opposition to euthanasia, he explains, may be based on this precept. We might be tempted to permit mercy killing when a dying person has only a short time to live; but Tendler performs a little moral arithmetic and concludes that 'a piece of infinity is also infinity, and a person who has but a few moments to live is no less of value than a person who has 60 years to live'.

Considered as a piece of reasoning, the logic of this is obviously defective. (One problem: the principle 'A piece of infinity is also infinity' is false. The series of natural numbers is infinitely long, but a piece of that series, such as the string 1–2–3–4–5, is not infinitely long.) But logical criticism seems almost irrelevant, because the formal argument seems little more than window-dressing for something much more fundamental. 'The infinite value of human life' is one of those suggestive slogans that expresses a deep, unreasoned conviction rather than a rational argument. What is going on here is an appeal to the idea that human life is *so* precious that the destruction of even a little bit of it is intolerable. The point is that if Mrs Rosier could have lived only one day longer, then her existence of those 24 hours would have been immensely valuable. On this view, the fact that those 24 hours would have had little value *for her* is irrelevant, for the conception of value being deployed here is not a conception of what is valuable for her. Her continued existence is taken to be good in some more abstract sense, perhaps as good for the universe. Indeed, it seems to be assumed that what is good for her should be sacrificed in the interest of this higher good which her mere existence constitutes.

The idea that human life has infinite value has another consequence. It also means that we can never compare lives and conclude that one is more valuable than another. Consider an infant with Tay-Sachs disease, who will inevitably die within a few months. Suppose we have only limited resources, and we must choose between caring for it, or caring instead for a healthy baby with a long life ahead of it. One might think that there is ample reason for preferring the latter. But, consistently with his slogan, Tendler thinks otherwise: because each and every human life has infinite value, he concludes that 'a handicapped individual is a perfect specimen when viewed in an ethical context. The value is an absolute value. It is not relative to life expectancy, to state of health, or to usefulness to society.'

Moral individualism, coupled with a more modest account of the value of human life, would handle such cases very differently. If we must choose between caring for the two babies, we would ask whether there are differences between them that would justify preferring one over the other. Such considerations as life expectancy and state of health would be relevant, because they determine the extent to which the infants' biological lives would be of use to them. Indeed, this choice would be easy, because the child with Tay-Sachs disease has no prospect of a biographical life at all, and so its biological life is of no value to it. That, on the more modest view, would be a decisive consideration.

It might be protested that this view leaves the value of human life less secure than traditional views. Indeed it does. The abandonment of lofty conceptions of human nature, and grandiose ideas about the place of humans in the scheme of things, inevitably diminishes our moral status. God and nature are powerful allies; losing them does mean losing something. But it does not mean losing everything. Human life can still be valued, and we can still justify moral and legal rules to protect it. We will, however, have to acknowledge that these rules grow out of our own valuings, rather than descending to us from some higher authority. If that is a loss, it may be a loss that humans after Darwin must live with.

At the same time, something is gained. Although it may seem odd to say so, in some respects traditional morality placed too much value on human life, and we might actually be better off with a more modest conception. Richard Schlatter, Patricia Rosier, and the Tay-Sachs infant are examples. Reverence for human life, which seems such a noble ideal in so many circumstances, can degenerate into a mindless superstition. When this happens, reason flies out of the window and what is in fact good for people is sacrificed to an abstract conception of their 'worth'. Avoiding these pitfalls is a considerable advantage of adopting a less grand view of our moral status.

The Right to Life

Some philosophers will think what I have said about the value of life is too anaemic, because, they will urge, I have taken no account of the *right* to life. When we say that people have a right to life, we are saying something more than merely that their lives are valuable. We are adding something extra. Intuitively, this seems correct. But it is not so easy to say exactly what this extra something is. One popular idea is that rights place absolute limits on what may be done to people. Suppose, for example, we could accomplish some great good by killing an innocent person—say, we could save the lives of five other innocent people. If we base our decision on what would lead to the best outcome, we might decide it is right to kill the one in order to save the five. But if, in killing this innocent person, we would be violating her *rights*, then we may not kill her no matter how much good would be accomplished thereby. Saving lives is a noble ideal, but we may not violate people's rights even in pursuit of noble ideals. Thus, on this view, rights are trumps that override all other considerations.

This view was popularized by Robert Nozick in *Anarchy, State, and Utopia*, in which he argued that rights are by their very nature inviolable. It is, however, an extremely strong conception of rights— perhaps too strong, for it implies that, if a person has a right to life, then it would be wrong to kill that person even if it was necessary to prevent the destruction of the whole world. As an alternative, I want to present a somewhat more modest account. We may begin by noticing that there is a difference, in our ordinary moral thinking, between (*a*) having a duty not to treat people in a certain way, and (*b*) violating their rights by treating them in that way. Then we ask what is the difference between these two cases. Three differences come immediately to mind. (We are, of course, talking throughout about moral rights, and not about legal rights, which involve somewhat different considerations.)

First, when an individual has a right not to be treated in a certain way, treating the person in that way is objectionable *for that individual's own sake*, and not merely for the sake of someone or something else. Suppose we are considering a Nazi concentration-camp 'experiment' in which all the subjects will be tormented and killed, and I say it would be wrong to treat Jones in that way. So far, so good. But suppose that my only objection is that Jones is an unsuitable subject and that if he is used the experiment might be spoiled—that is my only reason for thinking that using him would be wrong. Then I am not considering Jones himself to have any rights in the matter.

Secondly, when rights are involved, the rights-bearer is entitled to protest, in a special way, if he or she is not treated properly. Suppose you consider giving money for famine relief to be simply an act of generosity on your part. You think you ought to do it, because you ought to be generous to those in need. However, you do not consider the starving to have a *right* to your money; you do not owe it to them in any sense. Then, you will not think them entitled to complain if you choose not to give; it will not be proper for them to insist that you contribute, or to feel wronged if you do not. Whether you do your moral duty is in this case strictly between you and your conscience, and you owe them no explanation if you choose not to do so. On the other hand, if they have a *right* to your aid, it is permissible for them to complain, to insist, and to feel resentment if you do not give them what they have coming.

Thirdly, when rights are involved the positions of third parties are different. If you are violating someone's rights, it is permissible for a third party to intervene and compel you to stop. But if you are not violating anyone's rights, then even though you are not behaving as you ought, no third party is entitled to coerce you to do otherwise. Since giving for famine relief is widely considered not to involve the rights of the starving, but only to involve 'charity' towards them, it is not considered permissible for anyone to compel you to contribute. But if you had contracted to provide food, so that they now had a claim of right on your aid, compulsion would be thought proper if you reneged.

Not all philosophers are confident that moral rights even exist; some are suspicious of the concept and have wondered exactly what it means, and how it might be rendered in terms of the less puzzling notion of permissibility. The preceding observations suggest an analysis: X has a moral right to be treated in a certain way by Y if and only if, first, it is not permissible for Y not to treat X in that way, for reasons having to do with X's own interests; secondly, it is permissible for X to insist that Y treat him or her in that way, and to complain and feel resentment if Y does not; and thirdly, it is permissible for third parties to compel Y to treat X in that way if Y will not do so voluntarily. The *understanding* of rights that goes with this analysis is that rights are correlates of duties the performance of which we are not willing to leave to individual discretion.

On this account, then, to say that people have a right to life is just to say that they may not be killed, and that the reason has to do with their own interests; that they are entitled to insist on such respectful treatment; and that third parties may intervene to prevent their being killed. Plainly this is not the sort of absolute right that is implied by the Nozickian idea of rights-as-trumps. Nor does it require the introduction of any moral concepts beyond the ordinary notion of permissibility. It is, however, an idea of rights that is compatible with the modest account of the value of life we have sketched.

RETHINKING THE MORAL STATUS OF NON-HUMAN ANIMALS

One of the fundamental ideas expressed by moral individualism is that moral rules are species-neutral: the same rules that govern our treatment of humans should also govern our treatment of non-humans. We have already seen how this works with the principle of equality. But this idea is so profoundly contrary to what is usually assumed that it might be hard for some people even to take seriously, much less to accept. Therefore it may be useful to consider how it works with two other basic moral rules, the rule against killing and the rule against causing pain.

Killing. In the preceding section I outlined a theory of the value of human life and a corresponding account of the moral rule against killing. As a corollary, our theory should shed some light on the question of the value of non-human life. Once we have become clear about the reasons why killing humans is wrong, we are in a position to ask whether the same reasons, or similar ones, also apply in the case of non-humans.

Humans, we observed, are the subjects of lives—not just biological lives, but biographical lives. It is our lives in the biographical sense that we value, and the point of the rule against killing is to protect the interests that we have in virtue of the fact that we are the subjects of such lives. Do non-human animals also have biographical lives? Clearly, many do not. Having a life requires some fairly sophisticated mental capacities. Bugs and shrimp do not have those capacities. They are too simple. But consider a more complex animal such as the rhesus monkey. The rhesus is a favourite research animal for experimental psychologists because, being so close to us from an evolutionary point of view, they share many of our psychological characteristics. They are intelligent and live in organized social groups; they communicate with one another; they care about each other and, as we have seen, they behave altruistically towards one another. Monkey mothers and infants are bonded much as humans are. Moreover, they are not all alike: the lives and personalities of individual animals are surprisingly diverse. Their lives are not as intellectually and emotionally complex as those of humans, but clearly they do have lives.

Other examples could be given easily enough; twentieth-century investigators have confirmed Darwin's observation that the mental capacities of all the 'higher mammals' are similar to the capacities of humans. The situation seems to be that, when we consider the mammals closest to ourselves in the old phylogenetic scale, we find that they do have lives. Then the further down the scale we go, the less confidence we have that there is anything like a biographical life, until we reach the bugs and shrimp, where the notion of a biographical life has only the most doubtful application.

The moral view suggested by this is that animals, human and non-human, come under the protection of the rule against killing just to the extent that they are the subjects of biographical lives. But more needs to be said. There is no reason the wrongness of killing has to be an all-or-nothing matter; one killing could be more objectionable than another. Thus, killing an animal that has a rich biographical life might be more objectionable than killing one that has a simpler life. This corresponds fairly well to our pre-reflective intuitions. We think that killing a human is worse than killing a monkey, but we also think that killing a monkey is a more morally serious matter than squashing a bug. From an evolutionary perspective, this is fair enough. The lives of humans and non-humans need not be accorded exactly the same value, for the extra psychological capacities of humans provides reason why their lives may be valued more. At the same time, this does not mean that the lives of all other animals may be held cheap: on the contrary, consistency requires that to the extent that they have lives similar to our own, killing them must be regarded with a similar seriousness. The more complex their lives are, the greater the objection to destroying them.

This accords with some of our pre-reflective feelings, but it goes against others. The triumph of Darwinism during the past century has modified some of our intuitions about humans and other animals, but the transformation has by no means been complete. Our feelings are still largely shaped by pre-Darwinian notions. Thus many of us think nothing of killing even 'higher mammals' for food, to use their skins as ornamental clothing, or simply as sport. Moral individualism would require that these practices be reconsidered. Moreover, we feel instinctively that the life of *every* human being has what Kant called 'an intrinsic worth' or 'dignity' and so we tend to value every human life more than any non-human life, regardless of its particular characteristics. That is why the biological life of a Tay-Sachs infant, who will never develop into the subject of a biographical life, may be treated with greater respect than the life of an intelligent, sensitive animal such as a chimpanzee. Moral individualism would also imply that this judgement is mistaken.

Causing pain. Aquinas and Kant agreed that torturing animals is wrong, but they thought the reason has nothing to do with concern for the animals themselves. Rather, they said, torturing animals is wrong because it might lead one to be more cruel to humans. Moral individualism would reject such a view and say that cruelty to animals ought to be opposed, not merely because of the ancillary effects on humans, but because of the direct effects on the animals themselves. Animals that are tortured suffer, just as tortured humans suffer, and that is the primary reason it is wrong. In so far as both suffer, we have the *same* reason to oppose torturing one as the other, and it is inconsistent to take the one suffering but not the other as grounds for objection.

Although cruelty to animals is wrong, it does not follow that we are never justified in inflicting pain on an animal. Sometimes we are justified in doing this, just as we are sometimes justified in inflicting pain on humans. It does follow, however, that there must be a *good reason* for causing the suffering, and if the suffering is great, the justifying reason must be correspondingly powerful. As an example, consider the treatment of the civet cat, a highly intelligent and sociable animal. Civet cats are trapped and placed in small cages inside darkened sheds, where the temperature is kept up to 110° by fires. They are confined in this way until they die. What explains this extraordinary treatment? These animals have the misfortune to produce a substance that is useful in the manufacture of perfume. Musk, which is scraped from their genitals once a day for as long as they survive, makes the scent of perfume last a bit longer after each application. The heat affects their metabolism and increases their 'production' of musk, as does the scraping which involves deliberately keeping the genital area raw and swollen.

People are often shocked to hear this, and conclude immediately that the use of perfume made in this way is wrong. But it is worth considering exactly how the reasoning to this conclusion would go. The moral principle involved—the same one that applies to our treatment of humans—is that causing suffering is wrong unless there is a good reason to justify it. The production of perfume made with musk causes considerable suffering; the question, therefore, is whether our enjoyment of this product is a good enough reason to justify that suffering. Plainly it is not. Perfume itself is a trivial enough thing; but making the scent last a little longer is an even less significant concern. We could not conclude that our interest in this outweighs the animals' pain unless we thought that *any* human interest outweighs *every* non-human interest.

Giving up perfume made with musk, like giving up furs from animals caught with the infamous leg-hold trap, might be fairly easy. After all, those products are not very important to most of us. An exactly analogous argument can be given, however, in connection with the use of animals as food, and for most people this raises the possibility of a much more drastic alteration in their conduct. Animals that are raised and slaughtered for our consumption at mealtime also suffer, and the question is whether our enjoyment of the way they taste is a sufficient justification of their pain.

Most people radically underestimate the amount of suffering that is caused to animals raised and slaughtered for the table. They think, in a vague way, that slaughterhouses are cruel, and perhaps even that methods of slaughter ought to be made more humane. But after all, the visit to the slaughterhouse is a relatively brief episode in the animal's life; and beyond that, people imagine that the animals are treated well enough. But the truth is rather different.

Veal calves, for example, may spend their entire lives in pens too small to allow them to turn around or even to lie down comfortably— exercise toughens the muscles, which reduces the quality of the meat, and besides, allowing the animals adequate living space would be pro- hibitively expensive. In these pens the calves cannot perform such basic actions as grooming themselves, which they naturally desire to do, because there is not room for them to twist their heads around. Like human infants, calves want something to suck, and with their mothers unavailable they can be seen vainly trying to suck the sides of their stalls. In order to keep the meat pale and tasty, they are fed a liquid diet deficient in both iron and roughage. The calf's craving for iron becomes so intense that, if allowed to turn around, it will lick at its own urine, although calves normally find this repugnant. The tiny stall, which prevents the animal from turning, solves this problem.

Similar stories can be told about other animals on which we dine. In order to 'produce' animals by the millions, it is necessary to keep them crowded together in small spaces. Chickens are commonly kept eight or ten to a space smaller than a newspaper page. Unable to walk around or even stretch their wings—much less build a nest—the birds become vicious and attack one another. Among laying hens, the problem is sometimes exacerbated because the birds are so crowded that, unable to move, their feet grow around the wire floors of the cages, anchoring them to the spot. An anchored bird cannot escape attack no matter how desperately it tries. Mutilation of the animals is an efficient solution. To minimize the damage they can do to one another, the birds' beaks are cut off. The mutilation is painful (there are nerves in the beaks), but prob- ably not so painful as other sorts of mutilations that are routinely prac- tised. Cattle are castrated, not to prevent the unnatural 'vices' to which overcrowded chicken are prone, but because castrated cattle are more docile, put on more weight, and there is less risk of meat being 'tainted' by male hormones. Peter Singer reports that:

In Britain an anesthetic must be used, unless the animal is very young, but in America anesthetics are not in general use. The procedure is to pin the animal down, take a knife and slit the scrotum, exposing the testicles. You then grab each testicle in turn and pull on it, breaking the cord that attaches it; on older animals it may be necessary to cut the cord.

The moral rule against causing pain, once again, says that we may not cause suffering without a good reason, and if the amount of suffering is great, then the justifying reason must be correspondingly powerful. It is tempting to think that we do have a powerful reason for putting the animals through all this, namely, that we must do it in order to nourish ourselves. But only a little reflection is needed to see that this is not so. If we prefer to eat meat, it is not because we have to eat meat to survive—vegetarian meals are also nutritious. At bottom, our preference for a diet that includes meat is based simply on custom, and on the fact that we like the way the animals taste.

Does this mean we should stop eating meat? Such a conclusion will be resisted. 'What is objectionable', some will say, 'is not eating the animals, but only making them suffer. Perhaps we ought to protest about the way they are treated, and even work for better treatment of them. But it doesn't follow that we must stop eating them.' This sounds plausible until we realize that it would be impossible to treat the animals humanely and still produce meat in sufficient quantities to make it a normal part of our diets. Cruel methods are used in the meat-production industry not because the producers are cruel people, but because such methods are economical; they enable the producers to market a product that people can afford. So to work for better treatment of the animals would be to work for a situation in which most of us would have to adopt a vegetarian diet, because if we were successful we could no longer afford meat.

Vegetarianism is often regarded as an eccentric moral view, and it is assumed that a vegetarian must subscribe to principles at odds with common sense. But if this reasoning is sound, the opposite is true: the rule against causing unnecessary pain is the least eccentric of all moral principles, and that rule leads straight to the conclusion that we should abandon the business of meat production and adopt alternative diets. Considered in this light, vegetarianism might be thought of as a severely conservative moral stance.

Vivisection

Darwin's personal feelings about the mistreatment of animals were unusually strong, and matched in some ways his feelings about the mistreatment of humans. Reflecting on his father's character, Francis Darwin wrote: 'The two subjects which moved my father perhaps more strongly than any others were cruelty to animals and slavery. His detestation of both was intense, and his indignation was overpowering in case of any levity or want of feeling on these matters.' Numerous anecdotes illustrate the intensity of those feelings. Although he was generally mild-mannered and disliked public confrontation, Darwin could fly into a rage when he saw animals being abused. Francis reports that:

He returned one day from his walk pale and faint from having seen a horse ill-used, and from the agitation of violently remonstrating with the man. On another occasion he saw a horse-breaker teaching his son to ride, the little boy was frightened and the man was rough; my father stopped, and jumping out of the carriage reproved the man in no measured terms.

One other little incident may be mentioned, showing that his humanity to animals was well known in his own neighbourhood. A visitor, driving from Orpington to Down, told the cabman to go faster. 'Why', said the man, 'if I had whipped the horse *this* much, driving Mr Darwin, he would have got out of the carriage and abused me well.'

Darwin's feelings about animals were reflected in his pronouncement in *The Descent of Man* that 'humanity to the lower animals' is 'one of the noblest virtues with which man is endowed', and represents the final stage in the development of the moral sentiments. It is only when our concern has been 'extended to all sentient beings', he said, that our morality will have risen to its highest level.

Darwin was also involved in some public efforts to improve the treatment of animals. In 1863 he wrote an article for the *Gardener's Chronicle*, a popular monthly magazine, with the title 'Vermin and Traps'. 'The setting of steel traps for catching vermin', he argued, is too cruel a business for civilized people to tolerate. His rhetoric would not seem out of place in an animal-rights magazine today:

If we attempt to realise the sufferings of a cat, or other animal when caught, we must fancy what it would be to have a limb crushed during a whole long night, between the iron teeth of a trap, and with the agony increased by constant attempts to escape. Few men could endure to watch for five minutes, an animal struggling in a trap with a crushed and torn limb; yet on all the well-preserved estates throughout the kingdom, animals thus linger every night; and where game keepers are not humane, or have grown callous to the suffering constantly passing under their eyes, they have been known by an eyewitness to leave the traps unvisited for 24 or even 36 hours.

Darwin was careful to point out that the issue is not whether we feel sympathy for the animals, but only whether they suffer: 'We naturally feel more compassion for a timid and harmless animal,' he wrote, 'such as a rabbit, than for vermin, but the actual agony must be the same in all cases.' The animals' suffering has a direct, unmediated claim on our moral attention, and comparable suffering merits comparable concern regardless of whether the animal is timid, harmless, cute, or cuddly.

But as a man of science Darwin's moral views about animals were put to a severe test. In the 1870s what Francis Darwin called 'the anti-vivisection agitation' came to a boil in England. Public meetings were held, petitions were circulated, bills were introduced in Parliament, and scientists were thrown on the defensive. Darwin was inevitably drawn into the controversy. His humanitarian impulse collided with his desire to see science advance, and he was uncomfortably caught in the middle. Still, when pressed to choose, he chose science. His remarks on the subject show his discomfort. In 1871 he wrote to one correspondent,

You ask about my opinion on vivisection. I quite agree that it is justifiable for real investigations on physiology; but not for mere damnable and detestable curiosity. It is a subject which makes me sick with horror, so I will not say another word about it, else I shall not sleep tonight.

At times, we are told, visitors to Down House were forbidden to bring up the subject.

Despite these misgivings, Darwin eventually concluded that it was 'the duty of everyone whose opinion is worth anything [to] express his opinion publicly on vivisection'. In 1875 he testified before the Royal Commission on Vivisection, which included Huxley among its members, and took the lead, lobbying the Home Secretary, in trying to have a bill passed that would 'protect animals, and at the same time not injure physiology'. But a more radical bill was passed, which went further in protecting animals than Darwin thought wise.

The anti-vivisectionists were led by Francis Power Cobbe, one of those wonderful characters who seemed to be everywhere in Victorian England. 'Miss Cobbe', as she was called, was an ardent feminist from an early age. As a young woman in Dublin, she studied philosophy, wrote about Kant, and became a Unitarian preacher. In the early 1870s she founded the British Antivivisection Society as an alternative to the more conservative Royal Society for the Prevention of Cruelty to Animals. The RSPCA was an upper-class organization that focused its attention on abuses of animals among the lower classes; the common image invoked was that of a workingman beating his horse. Miss Cobbe's organization, however, was both more radical in its moral outlook and more egalitarian in its politics: its ire was aimed not at the workingmen but at upper-class scientists.

In her autobiography Miss Cobbe describes her friendly relationship with Darwin, which began when they were neighbours during the summer of 1869. Like others who knew him, she was struck by Darwin's tender feelings for animals:

He was glad to use a peaceful and beautiful old pony of my friend's yclept Geraint, which she placed at his disposal. His gentleness to this beast and incessant efforts to keep off the flies from his head, and his fondness for his dog Polly . . . were very pleasing traits in his character.

But Miss Cobbe's interests were not limited to animals. In 1869 Darwin was writing *The Descent of Man*, and they discussed the theory of moral development he was formulating. She tried to interest him in Kant's moral philosophy, but without much success—she insisted that he should read Kant; he demurred; and she sent him a copy of the *Grundlegung* anyway. Later they corresponded about the mental powers of dogs and about J. S. Mill's philosophy of science.

Remembering their conversation, Darwin sent Miss Cobbe a prepublication copy of *The Descent of Man* which, she declared, 'inspired me with the greatest alarm'. She wrote a review of the book for the *Theological Review* for April 1871. She realized that, in stressing the similarities between man and the animals, Darwin had provided her with a powerful argument for animal rights, and she later published an essay on 'Darwinism in Morals'. However, she was uneasy about the implications of evolution for human nature, and she never added evolutionism to her panoply of liberal enthusiasms. Eventually, though, the relationship between her and Darwin was broken off.

This pleasant intercourse with an illustrious man was, like many other pleasant things, brought to a close for me in 1875 by the beginning of the anti-vivisection crusade. Mr Darwin eventually became the centre of an adoring *clique* of vivisectors who (as his biography shows) plied him incessantly with encouragement to uphold their practice, till the deplorable spectacle was exhibited of a man who would not allow a fly to bite a pony's neck, standing forth before all Europe as the advocate of vivisection.

As part of the anti-vivisection campaign Cobbe wrote scorching letters to the London *Times*, and she was guilty, in Darwin's view, of unjustifiably villifying the scientists. He protested against 'the abuse poured in so atrocious a manner on all physiologists'. The members of the Royal Commission on Vivisection had concluded that British physiologists were not guilty of abusing their animal subjects, and Darwin accepted their judgement—although he thought European investigators were probably not so humane as their British counterparts. Still, the anti-vivisectionists' proposals, which Darwin rejected, were fairly mild. Miss Cobbe and her compatriots were not, by and large, abolitionists. They did not argue that all painful experiments on animals should be banned. Their goals were more modest than that. They wanted repetitious research eliminated, so that fewer animals would be needed. They asked that experimentation on live animals be limited to research that promised beneficial results, and that anaesthetics always be used. And finally, they favoured licensing so that private individuals would not be free to do whatever they pleased without being held accountable. Darwin's reasons for rejecting these seemingly modest proposals were explained in a letter to one of his daughters, who evidently had been pressing him on the subject:

Your letter has led me to think over vivisection (I wish some new word like anaes-section could be invented) for some hours, and I will jot down my conclusions, which will appear very unsatisfactory to you. I have long thought physiology one of the greatest of sciences, sure sooner, or more probably later, greatly to benefit mankind; but, judging from all other sciences, the benefits will accrue only indirectly in the search for abstract truth. It is certain that physiology can progress only by experiments on living animals. Therefore the proposal to limit research to points of which we can now see the bearings in regard to health, etc., I look at as puerile. I thought at first it would be good to limit vivisection to public laboratories; but I have heard only of those in London and Cambridge, and I think Oxford; but probably there may be a few others. Therefore only men living in a few great towns would carry on investigation, and this I should consider a great evil. If private men were permitted to work in their own houses, and required a license, I do not see who is to determine whether any particular man should receive one. It is young unknown men who are the most likely to do good work. I would gladly punish severely any one who operated on an animal not rendered insensible, if the experiment made this possible; but here again I do not see that a magistrate or jury could possibly determine such a point. Therefore I conclude, if (as is likely) some experiments have been tried too often, or anaesthetics have not been used when they should have been, the cure must be in the improvement of humanitarian feelings. Under this point of view I have rejoiced at the present agitation. If stringent laws are passed, and this is likely, seeing how unscientific the House of Commons is, and that the gentlemen of England are humane, as long as their sports are not considered, which entail a hundred or thousand-fold more suffering than the experiments of physiologists—if such laws are passed, the result will assuredly be that physiology, which has been until within the last few years at a standstill in England, will languish or quite cease. It will then be carried on solely on the Continent; and there will be so many the fewer workers on this grand subject, and this I should greatly regret.

Darwin did not object to the moral impulse behind the proposed rules— he seems to have approved of that—and, in a certain sense, he did not really object to the content of the rules. His concern seems to have been only with the consequences of their adoption, and with whether they could be administered intelligently. Today such concerns might seem to have less point. We are now accustomed to widespread government regulation of almost everything, including scientific research. Experience since Darwin's day shows that research can be regulated with a reasonable degree of intelligence and without its being caused to 'languish or quite cease'.

The intervening century has brought other changes that Darwin could hardly have dreamed possible. He wanted research to expand, not contract, and he would have been astounded by the extent to which his wish has been granted. In the United States alone, between 18 and 23 million animals are used in laboratory research annually. (This conservative estimate is provided by the Office of Technology Assessment, an agency of the US Congress; animal-rights groups give higher estimates, but no one knows for sure.) Darwin also wanted the moral debate about the use of these animals to be settled, in favour of science. He would be more disappointed in what has happened in this regard. The debate has continued, with the greater number of animals now involved making the argument even more intense.

Like the 1870s, the 1970s was a period of increased agitation on behalf of laboratory animals. This time the activists could muster weightier intellectual support than Miss Cobbe could manage. First the animal issue attracted the attention of the Australian philosopher Peter Singer, and in 1975, exactly 100 years after Darwin's letter to his daughter, he published *Animal Liberation*. Part philosophy and part activist tract ('The only philosophy book that contains recipes', observed one reader, referring to Singer's instructions for becoming a vegetarian), it became the handbook of a new and more aggressive animal-welfare movement. The scientific establishment, increasingly stung by the activities of militant organizations such as People for the Ethical Treatment of Animals, and often getting a bad press at the same time, was thrown on the defensive.

Then in 1982, Tom Regan's *Case for Animal Rights* took an even more radical stance. Regan, an American philosopher, argued that Singer's defence of animals did not go far enough. It is not enough, he said, to be concerned for animal *welfare*. Welfare is something that can be taken into account and then traded off against other values. Singer, a utilitarian, would acknowledge that if an experiment was designed so as to minimize suffering, and if it actually did more good than harm, it could be justified. He only criticized the great mass of research that could not pass even this minimum test. But Regan would have none of this utilitarian calculating. Instead, he said, we must acknowledge that, like humans, animals have *rights* that should not be violated under any circumstances whatever, not even if we think there is a great good to be achieved. While Singer was a reformer, Regan was an abolitionist.

To those uninterested in academic philosophy, however, the difference seemed small. Both were defenders of animals and both were critics of the scientific establishment. The appeal of the movement, whether it emphasized animal welfare or animal rights, depended not on such intellectual niceties but on the indignation people could be brought to feel about what was being done to the rats, dogs, and monkeys in the laboratories. Singer's book contained vivid descriptions. Scientists denounced them as misleading; but others, who had no idea what researchers actually do, found them eye-opening.

One series of experiments, which became infamous after Singer's book, was conducted by psychologists Harry F. Harlow and Stephen J. Suomi at the University of Wisconsin in the late 1960s. The use of animals in fields such as psychology is another twentieth-century development that Darwin could not have anticipated. For him, physiology was *the* science that needed animal subjects; but today physiologists use only a small percentage of the research animals. Millions are used by psychologists and by workers in altogether different areas such as the development and testing of commercial products.

Harlow and Suomi were interested in studying the psychopathology caused by maternal rejection, so they decided to take some rhesus monkey infants who had been rejected by their mothers and study the form their psychopathology would take. But immediately they encountered a problem: rhesus monkeys were chosen for the study because of their psychological resemblance to humans, and the investigators could not figure out how to get a monkey mother, whose attachment to her babies resembles that of human mothers, to reject her infant. The solution was to use a mechanical substitute to simulate maternal rejection: they would attempt to induce psychopathology 'by allowing baby monkeys to attach to cloth surrogate mothers who could become monsters'. This they described as 'a fascinating idea', but unfortunately it didn't work out as they had hoped. Here is Harlow's and Suomi's own description of their first efforts:

The first of these monsters was a cloth monkey mother who, upon schedule or demand, would eject high-pressure compressed air. It would blow the animal's skin practically off its body. What did this baby monkey do? It simply clung tighter and tighter to the mother, because a frightened infant clings to its mother at all costs. We did not achieve any psychopathology.

However, we did not give up. We built another surrogate monster mother that would rock so violently that the baby's head and teeth would rattle. All the baby did was cling tighter and tighter to the surrogate. The third monster we built had an embedded wire frame within its body which would spring forward and eject the infant from its ventral surface [i.e. its front]. The infant would subsequently pick itself off the floor, wait for the frame to return into the cloth body, and then again cling to the surrogate. Finally we built our porcupine mother. On command, this mother would eject sharp brass spikes over all of the ventral surfaces of its body. Although the infants were distressed by these pointed rebuffs, they simply waited until the spikes receded and then returned and clung to the mother.

The researchers note that 'These infant monkeys' behaviors were not surprising' because 'The only recourse of an injured or rebuked child—monkey or human—is to make intimate contact with the mother at any cost.'

It was then that Harlow and Suomi hit on the idea of creating a real monster mother by social isolation. For several years Harlow had been investigating the effects of social isolation on rhesus monkeys, using a small stainless-steel device called a 'vertical isolation chamber'. The idea for the chamber had been suggested to him by the fact that depressed people are sometimes described as 'sunken in a well of despair', and in one article he had written that the device 'was designed on an intuitive basis to reproduce such a well both physically and psychologically for monkey subjects'. Monkeys would be placed in the chamber a few hours after birth and kept there for up to eighteen months, in complete isolation, with nothing at all to do. Harlow found that 'sufficiently severe and enduring early isolation reduces these animals to a social-emotional level in which the primary social responsiveness is fear.'

Attempting to produce a real monster mother, Harlow and Suomi raised female monkeys in the vertical isolation chamber and then impregnated them with a device they called a 'rape rack'. (The potential mothers could not be allowed normal sexual relations with males because that would violate the conditions of isolation; and besides, it had already been learned that monkeys raised in isolation could make 'only ill-directed and infantile efforts at copulation'.) How would these mothers, reared in isolation and artificially impregnated, treat their babies? They turned out to be much more abusive than any of the surrogates the experimenters had created. Harlow and Suomi reported that:

They tended to show one of two syndromes. One pattern of the motherless mothers was to pay no attention to their infants. (Any normal monkey mother hearing one cry would have clasped the baby to its breast in no time flat.) The other mothers were brutal or lethal. One of their favorite tricks was to crush the infant's skull with their teeth. But the really sickening behavior pattern was that of smashing the infant's face to the floor, then rubbing it back and forth.

However, no psychopathology was ever produced in the babies, because even with the worst of the monster mothers, the babies never stopped coming back until either they were killed or the mothers began to show more normal maternal behaviour. So, in this respect at least, the experiments were a failure.

Was there anything morally objectionable about these experiments? Harlow and Suomi argued that they were trying to help accomplish something good. They were studying psychopathology, and their work, along with that of other investigators, could lead eventually to finding new forms of treatment for psychologically disturbed humans. But to accomplish anything they needed psychopathological subjects to study. They assumed, reasonably enough, that it would be unethical to induce psychopathology in humans; so they were doing the next-best thing, inducing psychopathology in beings as much like humans as possible.

But moral individualism would insist that, if it is wrong to use humans in experiments, then it is also wrong to use animals, unless there are relevant differences between them that justify a difference in treatment. Harlow and Suomi themselves provide a good bit of pertinent information about the animals used in their work. They go to considerable lengths in describing how similar to human children the baby monkeys are. They explain that maternal love is just as important to them as to human infants; indeed, they say that there is 'little difference' between the two kinds of babies in their emotional and intellectual needs—except that the baby monkeys are generally smarter. They also emphasize that normal monkey mothers are eager to provide support, and show the same sort of maternal affection as human mothers, and the same sort of distress when their babies are hurt. All this creates doubts about whether relevant differences can be found.

The problem may be expressed in the form of a dilemma that can arise for any psychological research that uses animals as models for the human case. If the animal subjects are not sufficiently like us to provide a model, the experiments may be pointless. (That is why Harlow and Suomi went to such lengths in stressing the similarities between humans and rhesus monkeys.) But if the animals are enough like us to provide a model, it may be impossible to justify treating them in ways we would not treat humans. The researchers are caught in a logical trap: in order to defend the usefulness of the research, they have to emphasize the similarities between the animals and the humans; but in order to defend it ethically, they must emphasize the differences. The problem is that one cannot have it both ways.

Darwin thought the sacrifice of animals to benefit humans is acceptable, but it is the triumph of his evolutionary viewpoint that makes the dilemma possible. The whole idea of using animals as psychological models for humans is a consequence of Darwinism. Before Darwin, no one could have taken seriously the thought that we might learn something about the human mind by studying mere animals. Similarly, the idea that we might object to mistreating animals *for the same reasons* that we object to mistreating humans is a distinctively post-Darwinian notion: it depends on not regarding humans and non-humans as fundamentally different. Considering his remarks about vivisection, it is fair to say that Darwin himself did not fully appreciate the implications of his own work.

CONCLUSION

As we have seen, Darwin believed that 'direct arguments' would have little effect on religious belief. Instead, he thought that 'the gradual illumination of men's minds', by the advancement of science, would result in the weakening of theism. There is obviously something to this. It is rare for anyone to change their religious views because of arguments; religion draws upon resources much too strong to be countered by mere ratiocination. The advancement of science, on the other hand, irresistibly alters one's whole picture of the way things are. It is a much more potent force. Something similar may be said about moral belief. It is almost as rare for anyone to change their moral views because of a mere argument. Even if every argument in this book were correct, it would be astonishing if readers simply accepted its conclusions.

Rather than being accomplished directly by arguments, we might think of moral change as the result of a more complicated historical process, in which arguments play a subordinate part. The particular process we have been considering has four stages. In the first stage, traditional morality is comfortably accepted because it is supported by a world-view in which everyone (or, so nearly everyone as makes no difference) has confidence. The moral view is deceptively simple. Human beings, as Kant put it, have 'an intrinsic worth, i.e., *dignity*', which makes them valuable 'above all price'; while other animals '. . . are there merely as means to an end. That end is man.' The world-view that supported this ethical doctrine had several familiar elements: the universe, with the earth at its centre, was seen as created by God primarily to provide a home for humans, who were made in his image; the other animals being created by God for their use. Humans, therefore, are set apart from other animals and have a radically different nature. This justifies their special moral standing.

In the second stage, the world-view begins to break up. This had begun to happen, of course, long before Darwin—it was already known that the earth is not the centre of the cosmos, and indeed, that considered as a celestial body it seems to be nothing special. But Darwin completed the job, by showing that humans, far from being set apart from the other animals, are part of the same natural order, and indeed, are actually kin to them. By the time Darwin was done, the old world-view was virtually demolished.

This did not mean, however, that the associated moral view would be immediately abandoned. Firmly established moral doctrines do not lose their grip overnight, sometimes not even overcentury. As Singer observes, 'If the foundations of an ideological position are knocked out from under it, new foundations will be found, or else the ideological position will just hang there, defying the logical equivalent of the law of gravity.'

We are now in the third stage, which comes when people realize that, having lost its foundations, the old moral view needs to be re-examined. In reviewing Tom Regan's book in defence of animal rights, Robert Nozick remarked that 'Nothing much should be inferred from our not presently having a theory of the moral importance of species member- ship that no one has spent much time trying to formulate because the issue hasn't seemed pressing.' The issue hasn't seemed pressing because philosophers have not yet fully assimilated the implications of the collapse of the old world-view.

It still might turn out that traditional morality is defensible, if new support can be found for it. Nozick, and a host of others, think this is likely. This book has argued otherwise: 'the gradual illumination of men's minds' must lead to a new ethic, in which species membership is seen as relatively unimportant. The most defensible view seems to be some form of moral individualism, according to which what matters is the individual characteristics of organisms, and not the classes to which they are assigned. The heart of moral individualism is an equal concern for the welfare of all beings, with distinctions made among them only when there are relevant differences that justify differences in treatment. It may be that there is some better view consistent with the spirit of Darwinism. I do not now think so; but it is always unwise to assume complacently that one has found The Truth. Be that as it may, the issues pressed upon us by the disintegration of the old world-view can no longer be avoided. The fourth and final stage of the process will be reached if and when those issues are resolved, and a new equilibrium is found in which our morality can once again comfortably coexist with our understanding of the world and our place in it.

Darwin had his own view of the direction that moral progress might take. As we have seen, he believed that our moral sentiments must eventually expand to include all mankind, regardless of nation, race, social status, or handicap, 'and finally the lower animals'. It is tempting to regard these moral pronouncements as philosophical fancy, noble in themselves, but overly idealistic and in any case unrelated to his strictly scientific achievement. But there is another possibility, which I have tried to defend in this book: that Darwin was correct in thinking that all his work, from the theory of natural selection to the moral vision he articulates, is of one piece. It is one view, held together by a sense of how the elements of one's thinking must be mutually supportive, and how they must fit together, if one's outlook is to form a reasonable and satisfying whole.

ACKNOWLEDGEMENTS

In thinking about the subjects dealt with in this book, I have benefited from the work of other writers in ways that are not always apparent. Therefore, I want to mention and give credit to some of the works that I have found especially useful.

Stephen Jay Gould's essays about Darwin and other figures in nineteenth-century science are models of good writing and sensitive interpretation. The fact that I disagree with Gould about central philosophical issues—he doesn't think that Darwinism has any religious or moral implications—should not obscure the fact that I have learned a great deal from his writings.

For miscellaneous information about Darwin's life, I found Ronald Clark's *Survival of Charles Darwin* to be especially useful. His book *The Huxleys* was equally helpful regarding T. H. Huxley. For information about Asa Gray, I relied on A. Hunter Dupree's excellent biography.

David Hull's long introductory essay to *Darwin and His Critics* is a gem. I especially benefited from his discussion of teleology. On the ways in which modern science differs from Aristotelian science, I have learned much from Everett Hall's *Modern Science and Human Values* and E. M. Adams' *Ethical Naturalism and the Modern World-View*. My remarks about preformationism are based on the account given by Richard Westfall in his *Construction of Modern Science*.

In what I say concerning the independence of ethics from a sociobiological account of the origins of conduct, I was influenced by Thomas Nagel's argument in his 'Ethics as an Autonomous Theoretical Subject'. The general account of sociobiology from which I have learned the most is Philip Kitcher's *Vaulting Ambition*.

'Anthropocentrism: Bad Practice, Honest Prejudice?' by D. R. Crocker, gave me the idea of contrasting what ethologists say to the public with what they say to one another. Some of the examples I cite were suggested by Crocker also.

Finally, four friends have contributed to this book in equally direct ways. I became interested in writing about Darwin while jointly teaching a course on 'Darwin, Marx, and Freud' at UAB with Theodore M.

Benditt. Talking with Ted Benditt about Darwin was the First Cause of this book. Later, my colleague Harold Kincaid, whose knowledge of the philosophy of science far exceeds my own, helped me with a number of points. Later still, E. Culpepper Clark and Stuart Rachels separately read a draft of the book and made helpful comments about both style and substance. To these four people I am especially grateful.

REFERENCES

THE following list identifies the sources of all quotations in this book. Full information about the works cited is given in the Bibliography. References to *The Origin of Species* and *The Descent of Man* are to the first editions of these works, unless otherwise noted. In some instances I have altered spelling and punctuation. For example, in his letters and notebooks Darwin habitually shortened 'could' to 'c^d' and 'should' to 'sh^d'. In quoting him, I use the words, not the abbreviations.

Introduction

'Man in his arrogance . . .' Darwin, *Notebooks*, 300.
'The Darwinian theory . . .' Wittgenstein, *Tractatus*, 49.
'Darwin's book is very important . . .' Marx, quoted in Zirkle, *Evolution, Marxian Biology, and the Social Scene*, 86.
'accept and welcome . . .' Carnegie, *The Gospel of Wealth*, 399.
'While I'm not a conventional believer . . .' Gould, 'Darwinism Defined', 70.
'What challenge can the facts . . .' Gould, 'Darwinism Defined', 70.

Chapter 1: Darwin's Discovery

'You care for nothing . . .' Darwin, *Autobiography*, 28.
'Darwin had a youth . . .' Clark, *The Survival of Charles Darwin*, 6.
'One day, on tearing off . . .' Darwin, *Autobiography*, 62.
'Caught a sea-mouse . . .' Quoted in Clark, *The Survival of Charles Darwin*, 10.
'Mr Darwin communicated . . .' Quoted in Clark, *The Survival of Charles Darwin*, 11.
'sufficient to check . . .' Darwin, *Autobiography*, 46
'I did not then . . .' Darwin, *Autobiography*, 57.
'From my passion . . .' Darwin, *Autobiography*, 60.
'I know that I ought . . .' Darwin, *Autobiography*, 60.
'knew every branch of science' Darwin, *Autobiography*, 64.
'I was called . . .' Darwin, *Autobiography*, 64.
'Capt F. wants . . .' Darwin and Henslow, *Darwin and Henslow*, 30.
'The marks of *design* . . .' Paley, *Evidences*, 473.
'A scheme of nature . . .' Huxley, *Science and Culture*, 321.
'the proportion of speculation . . .' Darwin, *Autobiography*, 49.

'wonderfully superior' Darwin, *Autobiography*, 77.

'One day . . .' Darwin, *Autobiography*, 49.

'I never before . . .' Darwin, *Correspondence*, i. 226.

'FitzRoy's character . . .' Darwin, *Autobiography*, 72–3.

'We had several quarrels . . .' Darwin, *Autobiography*, 73–4.

'a man who has . . .' Darwin, *Correspondence*, ii. 80.

'The day has passed . . .' Darwin, *Journal of Researches*, 11–12.

'an Englishman . . .' Darwin, *Journal of Researches*, 19.

'The minute I landed . . .' Darwin, *Correspondence*, i. 342.

'doing admirable work . . .' Sedgwick, *Life and Letters*, i. 380.

'I frequently got on . . .' Darwin, *Journal of Researches*, 385.

'I have not as yet noticed . . .' Darwin, *Journal of Researches*, 394.

'Seeing this gradation . . .' Darwin, *Journal of Researches*, 381.

'those singular rings . . .' Darwin, *The Structure and Distribution of Coral Reefs*, 88.

'the whole theory . . .' Darwin, *Autobiography*, 98.

'I have always felt . . .' Darwin, *Autobiography*, 77.

'I may mention one . . .' Darwin, *Journal of Researches*, 24–5.

'And think of Darwin's position . . .' Gould, *Ever Since Darwin*, 33.

'In October 1838 . . .' Darwin, *Autobiography*, 120.

'It is impossible to say . . .' Hooker, *Life and Letters*, ii. 299.

'three stages . . .' Hooker, *Life and Letters*, ii. 299.

'secretion of the brain' Darwin, *Notebooks*, 291.

'In virtually every branch . . .' Gruber, *Darwin and Man*, 203.

'If Wallace had my MS sketch . . .' Darwin, *Life and Letters*, i. 473.

'I can plainly see . . .' Darwin, *Life and Letters*, i. 453–4.

'I believe I go much further . . .' Darwin, *Life and Letters*, i. 466.

'I shall, of course . . .' Darwin, *Life and Letter*, i. 473.

'it would be dishonourable . . .' Darwin, *Life and Letters*, i. 475.

'I admire extremely . . .' Darwin, *Life and Letters*, i. 500.

'I have received . . .' Wallace, *Letters and Reminiscences*, ii. 57.

'As to the theory . . .' Wallace, *Letters and Reminiscences*, ii. 131.

'Some of my critics . . .' Darwin, *Autobiography*, 140.

'How extremely stupid . . .' Huxley, 'On the Reception of "The Origin of Species"' in Darwin, *Life and Letters*, ii. 197.

'The key is man's power . . .' Darwin, *Origin*, 30.

'No doubt the strawberry . . .' Darwin, *Origin*, 41.

'we cannot recognize . . .' Darwin, *Origin*, 37.

'Breeders habitually speak . . .' Darwin, *Origin*, 31.

'differences which I for one . . .' Darwin, *Origin*, 32.

'follow from the struggle . . .' Darwin, *Origin*, 61.

'Here we see how potent . . .' Darwin, *Origin*, 71.

'Hence, it is quite credible . . .' Darwin, *Origin*, 74.

'Grouse, if not destroyed . . .' Darwin, *Origin*, 84.

'yet we hear . . .' Darwin, *Origin*, 85.

'Natural Selection has been . . .' Darwin, *Origin*, 2nd edn., 14.

'Breeders believe that long limbs . . .' Darwin, *Origin*, 11.

'This form of selection . . .' Darwin, *Origin*, 2nd edn., 69.

'A hornless stag . . .' Darwin, *Origin*, 2nd edn., 69.

'I can see no good reason . . .' Darwin, *Origin*, 2nd edn., 69.

'Much light will be thrown . . .' Darwin, *Origin*, 488.

'Now, we must say . . .' Wilberforce, quoted in Clark, *The Survival of Charles Darwin*, 145.

'the most formidable . . .' Eiseley, *Darwin and the Mysterious Mr X*, 13.

'The Lord hath delivered him . . .' Huxley supposedly said this to Sir Benjamin Brodie, surgeon to the Queen.

'Many blamed Huxley . . .' Lyell, *Life, Letters, and Journals*, ii. 335.

'I knew of this theory . . .' Hooker, quoted in Clark, *The Survival of Charles Darwin*, 144.

'The Bishop had been much applauded . . .' Lyell, *Life, Letters, and Journals*, ii. 335.

'I have read your book . . .' Sedgwick, *Life and Letters*, ii. 356.

'Poor dear old Sedgwick . . .' Darwin, *Life and Letters*, ii. 200.

'If the book be true . . .' Sedgwick, *Life and Letters*, ii. 82.

'all our morality . . .' Lyell, *Life, Letters, and Journals*, i. 186.

'Whether the naturalist believes . . .' Darwin, *Collected Papers*, ii. 81.

'the structural differences . . .' Huxley, *Man's Place in Nature*, 123.

'If this work had appeared . . .' Darwin, *Descent of Man*, 4.

'every chief fissure and fold . . .' Darwin, *Descent of Man*, 10–11.

'The power of erecting . . .' Darwin, *Descent of Man*, 2nd edn., 402.

'As soon as some . . .' Darwin, *Descent of Man*, 2nd edn., 433–7.

'a secretion of the brain' Darwin, *Notebooks*, 291.

'There is no fundamental . . .' Darwin, *Descent of Man*, 35.

'I cannot, therefore . . .' Darwin, *Descent of Man*, 2nd edn., 432.

'the older and honoured chiefs' Darwin, *Descent of Man*, 2nd edn., 389.

Chapter 2: How Evolution and Ethics Might be Related

'I have noted . . .' Quoted in Clark, *The Survival of Charles Darwin*, 205.

'It has also pleased me . . .' Darwin, *Life and Letters*, iii. 120.

'Mr Spencer represents . . .' Hofstadter, *Social Darwinism in American Thought*, 33.

'We have in Herbert Spencer . . .' Hofstadter, *Social Darwinism in American Thought*, 31.

'The growth of large business . . .' Hofstadter, *Social Darwinism in American Thought*, 45.

'While the law . . .' Hofstadter, *Social Darwinism in American Thought*, 46.

'the establishment of rules . . .' Spencer, *The Data of Ethics*, iv.

'Now that moral injunctions . . .' Spencer, *The Data of Ethics*, iv.

'has for its subject-matter . . .' Spencer, *The Data of Ethics*, 21.

'permanently peaceful . . .' Spencer, *The Data of Ethics*, 20.

'the limit of evolution . . .' Spencer, *The Data of Ethics*, 19–20.

'The conduct to which . . .' Spencer, *The Data of Ethics*, 26–7.

'In every system . . .' Hume, *A Treatise of Human Nature*, 469.

'Thus there is no escape . . .' Spencer, *The Data of Ethics*, 30.

'The *Origin of Species* introduced . . .' Dewey, *The Influence of Darwin on Philosophy*, 1.

'O my Bergson . . .' Perry, *The Thought and Character of William James*, ii. 618.

'the religious liberals . . .' Bergson, *Creative Evolution*, x.

'the systematic study . . .' Wilson, *Sociobiology*, 595.

'The populace . . .' Wilson, *Sociobiology*, 553.

'In hunter-gatherer societies . . .' Wilson, 'Human Decency is Animal'.

'biological basis . . .' Barash, *The Whisperings Within*, 3.

'kow-towing to feminism' van den Berghe, *Human Family Systems*, 2.

'The range . . .' Gould, *Ever Since Darwin*, 252.

'What is the direct evidence . . .' Gould, *Ever Since Darwin*, 254.

'ethical philosophers who wish . . .' Wilson, *Sociobiology*, 3.

'The time has come . . .' Wilson, *Sociobiology*, 562.

'a theoretical inquiry . . .' Nagel, 'Ethics as an Autonomous Theoretical Subject', 196.

'You may well believe . . .' Lyell, *Life, Letters, and Journals*, ii. 376.

'I am sick and tired . . .' Clark, *The Huxleys*, 44.

'Oh, no, Professor . . .' Clark, *The Huxleys*, 44.

'I trust you will not . . .' Clark, *The Huxleys*, 51.

'On all sides I shall hear . . .' Huxley, *Man's Place in Nature*, 129.

'I have endeavoured . . .' Huxley, *Man's Place in Nature*, 129–30.

'Our reverence . . .' Huxley, *Man's Place in Nature*, 132.

'difficult and delicate matter' Gray, *Letters*, ii. 699.

'I thought you would . . .' Darwin, *Life and Letters*, ii. 477.

'A being who . . .' Gray, *Natural Science and Religion*, 103.

'May it not well be . . .' Gray, *Natural Science and Religion*, 105–6.

'Man, while on the one side . . .' Gray, *Natural Science and Religion*, 54.

'Of all parts . . .' Aquinas, *Basic Writings*, ii. 221.

'Christians have . . .' Augustine, *Basic Writings*, ii. 27.

'the essence of his soul' Augustine, *The City of God*, 152.

'contrary to that charity . . .' Aquinas, *Summa Theologica*, II–II, Q. 64, Art. 5.

'ends in themselves' Kant, *Foundations of the Metaphysics of Morals*, 46–7.

'an intrinsic worth . . .' Kant, *Foundations of the Metaphysics of Morals*, 53.

'Humanity is worthy . . .' Kant, *Lectures on Ethics*, 151.

'If [a man] disposes . . .' Kant, *Lectures on Ethics*, 151.

'The rule of morality . . .' Kant, *Lectures on Ethics*, 152.

'But as soon . . .' Kant, *Lectures on Ethics*, 153–4.

'Other creatures . . .' Aquinas, *Summa Contra Gentiles*, III, II, 112.

'The love of charity . . .' Aquinas, *Summa Theologica*, II–II, Q. 25. Art. 3.

'But so far as animals . . .' Kant, *Lectures on Ethics*, 239.

'He who is cruel . . .' Kant, *Lectures on Ethics*, 240.

'Every creature proclaims . . .' Quoted in Singer, *Animal Liberation*, 215.

'Francis had little . . .' Passmore, *Man's Responsibility for Nature*, 112.

'This small attention . . .' Hume, *A Treatise of Human Nature*, 470.

'Hume's Guillotine' Black, 'The Gap between "Is" and "Should"'.

'a stretching or growth . . .' Swammerdam, quoted in Westfall, *The Construction of Modern Science*, 100.

'An egg is . . .' Harvey, quoted in Westfall, *The Construction of Modern Science*, 97.

Chapter 3: Must a Darwinian be Sceptical about Religion?

'Design by wholesale . . .' Beecher, *Evolution and Religion*, 51.

'The teaching of the Church . . .' Pius XII, *Humani Generis*, 30.

'Most scientists . . .' Gould, 'Darwinism Defined', 70.

'Unless at least half . . .' Gould, 'Darwinism Defined', 70.

'I did not then . . .' Darwin, *Autobiography*, 57.

'Disbelief crept over me . . .' Darwin, *Autobiography*, 87.

'Though I am a strong advocate . . .' From a letter written by Darwin, supposedly to Karl Marx; but probably written to Marx's son-in-law Edward Aveling: published in Lucas, 'Marx und Engels'.

'This argument would be . . .' Darwin, *Autobiography*, 91.

'I can indeed hardly see . . .' Darwin, *Autobiography*, 87.

'That there is much suffering . . .' Darwin, *Autobiography*, 90.

'The presence of much suffering . . .' Darwin, *Autobiography*, 90.

'But pain or suffering . . .' Darwin, *Autobiography*, 89.

'that cause we call God' Aquinas, *Basic Writings*, i. 22.

'I had no intention . . .' Darwin, *Life and Letters*, ii. 105.

'I may say that . . .' Darwin, *Life and Letters*, i. 276.

'Disbelief crept over me . . .' Darwin, *Autobiography*, 87 (emphasis added).

'He naturally shrank . . .' Darwin, *Life and Letters*, i. 275.

'When I am dead . . .' Darwin, *Autobiography*, 237.

'Can the mind of man . . .' Darwin, *Autobiography*, 93.

'When I reached . . .' Huxley, *Science and Christian Tradition*, 237–9

'In my most extreme . . .' Darwin, *Life and Letters*, i. 274.

'Not only . . .' Marx, Letter to Lassalle, quoted in Clark, *The Survival of Charles Darwin*, 212.

'It is a mistake . . .' Nagel, *The Structure of Science*, 24.

'Nature belongs to the class . . .' Aristotle, *Basic Works*, 249.

'not for the sake of something . . .' Aristotle, *Basic Works*, 249.

'The old argument . . .' Darwin, *Autobiography*, 87.

'As far as the examination . . .' Paley, *Evidences*, repr. in Edwards and Pap, 425.

'Besides that conformity . . .' Paley, *Evidence*, repr. in Edwards and Pap, 427–8.

'To suppose that the eye . . .' Darwin, *Origin*, 186.

'In living bodies . . .' Darwin, *Origin*, 189.

'If the paleontological evidence . . .' Mavrodes, '"Creation Science" and Evolution', 43.

'Those who can . . .' Darwin, *On the Various Contrivances by which Orchids are Fertilised by Insects*, 261.

'The regular course . . .' Darwin, *On the Various Contrivances by which Orchids are Fertilised by Insects*, 282.

'If a man . . .' Darwin, *On the Various Contrivances by which Orchids are Fertilised by Insects*, 283–4.

'an excretion . . .' Darwin, *On the Various Contrivances by which Orchids are Fertilised by Insects*, 266.

'The larvae . . .' Darwin, *On the Various Contrivances by which Orchids are Fertilised by Insects*, 266.

'a divine direction . . .' Mavrodes, '"Creation Science" and Evolution', 43.

'An omniscient Creator . . .' Darwin, *The Variation of Animals and Plants under Domestication*, ii. 414–5.

'In reality these . . .' Freud, *The Future of an Illusion*, 32.

'doesn't intersect . . .' Gould, 'Darwinism Defined', 70.

Chapter 4: How Different are Humans from Other Animals?

'Charity does not extend . . .' Aquinas, *Summa Theologica*, II, Q. 65, Art. 3.

'My opinion is not . . .' Descartes, Letter to Henry More (1649), quoted in Singer, *Animal Liberation*, 219.

'They administered beatings . . .' Nicholas Fontaine, *Memoires*, quoted in Singer, *Animal Liberation*, 220.

'Researchers said . . .' *Birmingham News*, 2 Feb. 1986, 23A.

'To the best of our knowledge . . .' Lt. Col. William J. McGee, quoted in *Time*, 6 Feb. 1978, 50.

'Animals whom we have made . . .' Darwin, *Metaphysics, Materialism, and the Evolution of Mind*, 187.

'animated machines' Darwin, *Descent of Man*, 48.

'There is no fundamental . . .' Darwin, *Descent of Man*, 35.

'Of all the faculties . . .' Darwin, *Descent of Man*, 46.

'So many facts . . .' Darwin, *Descent of Man*, 47.

'Some animals extremely low . . .' Darwin, *Descent of Man*, 46.

'a result which has surprised . . .' Darwin, *Formation of Vegetable Mould*, 35.

'As I was led . . .' Darwin, *Formation of Vegetable Mould*, 2–3.

'not only to serve . . .' Darwin, *Formation of Vegetable Mould*, 58.

'One alternative alone . . .' Darwin, *Formation of Vegetable Mould*, 98.

'The sphex had not . . .' Darwin, *Formation of Vegetable Mould*, 94.

'strike every one . . .' Darwin, *Formation of Vegetable Mould*, 98.

'If worms have . . .' Darwin, *Formation of Vegetable Mould*, 97.

'These Victorian ladies . . .' Ferry, *The Understanding of Animals*, 2.

'I cannot doubt that language . . .' Darwin, *Descent of Man*, 56–7.

'Any one who has watched . . .' Darwin, *Expression of the Emotions*, 60.

'makes infinite use of finite means' Wilhelm von Humboldt, quoted in Chomsky, *Aspects of the Theory of Syntax*, v.

'By [this method] we may . . .' Descartes, *Philosophical Works*, i. 116–17.

'It is not credible . . .' Descartes, *Philosophical Works*, i. 117.

'As the voice was used . . .' Darwin, *Descent of Man*, 57.

'A long and complex train . . .' Darwin, *Descent of Man*, 57.

'Forget the use of language . . .' Gruber, *Darwin on Man*, 296.

'The orang in the Eastern islands . . .' Darwin, *Descent of Man*, 36.

'In a demonstration . . .' Skinner, 'Behaviorism at Fifty', 90.

'The pigeon *observed* . . .' Skinner, 'Behaviorism at Fifty', 91.

'The pigeon *felt* . . .' Skinner, 'Behaviorism at Fifty', 91.

'an automatic response . . .' Wooldridge, *The Machinery of the Brain*, 76.

'When the time comes . . .' Wooldridge, *The Machinery of the Brain*, 82–3.

'There will always be room . . .' Dennett, *Brainstorms*, 245.

'Even the lowest type . . .' Descartes, *Philosophical Works*, i. 116 (emphasis added).

'Of all the differences . . .' Darwin, *Descent of Man*, 70.

'The moral sense . . .' Darwin, *Descent of Man*, 97–8.

'Everyone must have noticed . . .' Darwin, *Descent of Man*, 74.

'Social animals perform . . .' Darwin, *Descent of Man*, 74–5.

'Capt Stansbury found . . .' Darwin, *Descent of Man*, 77.

'Brehm encountered . . .' Darwin, *Descent of Man*, 75–6.

'a majority of rhesus monkeys . . .' Masserman *et al.*, '"Altruistic" Behavior in Rhesus Monkeys', 585.

'were paired against each other . . .' Wechkin *et al.*, 'Shock to a Conspecific as an Aversive Stimulus', 47.

'the SAs vocalized infrequently' Wechkin *et al.*, 'Shock to a Conspecific as an Aversive Stimulus', 48.

'the rage and attack mimetics . . .' Masserman *et al.*, '"Altruistic" Behavior in Rhesus Monkeys', 585.

'This behavior of the shocked Os . . .' Wechkin *et al.*, 'Shock to a Conspecific as an Aversive Stimulus', 48.

'But it may be asked . . .' Darwin, *Descent of Man*, 163.

'With respect to the origin . . .' Darwin, *Descent of Man*, 80–1.

'In the first place . . .' Darwin, *Descent of Man*, 163–4.

'from the side of natural history' Darwin, *Descent of Man*, 71.

'the foundation of morality . . .' Darwin, *Descent of Man*, 97.

'the Greatest Happiness principle' Darwin, *Descent of Man*, 97.

'the general good . . .' Darwin, *Descent of Man*, 98.

'As the social instincts . . .' Darwin, *Descent of Man*, 98.

'Any animal whatever . . .' Darwin, *Descent of Man*, 71–2.

'Savages, for instance . . .' Darwin, *Descent of Man*, 97.

'The social instincts which . . .' Darwin, *Descent of Man*, 103.

'Why does man regret . . .' Darwin, *Descent of Man*, 89.

'Butler and MacIntosh . . .' Darwin, *Notebooks*, 628.

'Thus, as man cannot prevent . . .' Darwin, *Descent of Man*, 90.

'If he has no such sympathy . . .' Darwin, *Descent of Man*, 92.

'As man advances . . .' Darwin, *Descent of Man*, 100–1.

'his sympathies became more tender . . .' Darwin, *Descent of Man*, 103.

'can indeed solve . . .' Harlow and Harlow, *Lessons from Animal Behavior for the Clinician*, ch. 5; quoted in Godlovitch *et al.*, *Animals, Men, and Morals*, 75.

'A very capable student . . .' Fossey, *Gorillas in the Mist*, 56.

'Ruby's face grimaced . . .' Mackinnon, *In Search of the Red Ape*, 176.

'Due to the small size . . .' Mackinnon, 'The Behavior and Ecology of Wild Orang-utans', 57.

'As time progressed . . .' Laidler, *The Talking Ape*, 146.

'Cody very rarely . . .' Laidler, 'Language in the Orang-utan', 152.

'You think I humanize . . .' Lorenz, *King Solomon's Ring*, 152.

Chapter 5: Morality without the Idea that Humans are Special

'The shockwaves . . .' Mayr, *The Growth of Biological Thought*, 438–9.

'the ability to regulate . . .' Nozick, *Anarchy, State and Utopia*, 49.

'The racist violates . . .' Singer, *Animal Liberation*, 9.

'perhaps it will turn out . . .' Nozick, 'About Mammals and People', 29.

'Persons who are unable . . .' Cohen, 'The Case for the Use of Animals in Biomedical Research', 866 (emphasis added).

'To make covenants . . .' Hobbes, *Leviathan*, 90.

'We use the characterization . . .' Rawls, *A Theory of Justice*, 505.

'I look at the term species . . .' Darwin, *Origin*, 52.

'suicide is . . .' Kant, *Lectures on Ethics*, 151.

'Its advocates argue thus . . .' Kant, *Lectures on Ethics*, 148.

'past any sensible point' Erlanger, 'A Scholar and Suicide'.

'Man is not a thing . . .' Kant, *Lectures on Ethics*, 151–2.

'Human life is . . .' Moshe Tendler, quoted in Kuhse, *The Sanctity-of-Life Doctrine in Medicine*, 12.

'a handicapped individual . . .' Tendler, quoted in Kuhse, *The Sanctity-of-Life Doctrine in Medicine*, 12.

'In Britain an anesthetic . . .' Singer, *Animal Liberation*, 152.

'The two subjects . . .' Francis Darwin, quoted in Clark, *The Survival of Charles Darwin*, 76.

'He returned one day . . .' Darwin, *Life and Letters*, iii. 200.

'humanity to the lower animals' Darwin, *Descent of Man*, 101.

'The setting of steel traps . . .' Darwin, *Collected Papers*, ii. 83.

'If we attempt to realise . . .' Darwin, *Collected Papers*, ii. 83–4.

'We naturally feel more compassion . . .' Darwin, *Collected Papers*, ii. 84.

'the anti-vivisection agitation' Darwin, *Life and Letters*, iii. 201.

'You ask about my opinion . . .' Darwin, *Life and Letters*, iii. 200.

'the duty of everyone . . .' Darwin, *Life and Letters*, iii. 209.

'protect animals . . .' Darwin, *Life and Letters*, iii. 204.

'He was glad to use . . .' Cobbe, *Life*, ii. 445.

'inspired me with the greatest alarm' Cobbe, *Life*, ii. 447.

'This pleasant intercourse . . .' Cobbe, *Life*, ii. 449.

'the abuse poured . . .' Darwin, *Life and Letters*, iii. 206.

'Your letter has led me . . .' Darwin, *Life and Letters*, iii. 202–3.

'by allowing baby monkeys . . .' Harlow and Soumi, 'Induced Psychopathology in Monkeys', 9.

'The first of these . . .' Harlow and Soumi, 'Induced Psychopathology in Monkeys', 9.

'These infant monkeys . . .' Harlow and Soumi, 'Induced Psychopathology in Monkeys', 9.

'sunken in a well . . .' Soumi and Harlow, 'Depressive Behavior in Young Monkeys', 11.

'was designed . . .' Soumi and Harlow, 'Depressive Behavior in Young Monkeys', 11.

'sufficiently severe . . .' Harlow, Doddsworth, and Harlow, 'Total Isolation in Monkeys', 90.

'only ill-directed . . .' Harlow and Soumi, 'Induced Psychopathology in Monkeys', 10.

'They tended to show . . .' Harlow and Soumi, 'Induced Psychopathology in Monkeys', 10.

'If the foundations . . .' Singer, *Animal Liberation*, 231.

'Nothing much . . .' Nozick, 'About Mammals and People', 29.

BIBLIOGRAPHY

ADAMS, E. M., *Ethical Naturalism and the Modern World-View* (Chapel Hill: University of North Carolina Press, 1960).

AQUINAS, ST THOMAS, *Basic Writings of St Thomas Aquinas*, ed. Anton C. Pegis (2 vols.; New York: Random House, 1945).

—— *Summa Contra Gentiles*, trans. English Dominican Fathers (New York: Benziger Brothers, 1928).

—— *Summa Theologica*, trans. English Dominican Fathers (New York: Benziger Brothers, 1918).

ARDREY, ROBERT, *The Territorial Imperative* (New York: Atheneum, 1966).

ARISTOTLE, *The Basic Works of Aristotle*, ed. Richard McKeon (New York: Random House, 1941).

AUGUSTINE, ST, *Basic Writings of Saint Augustine*, ed. Whitney J. Oates (2 vols.; New York: Random House, 1948).

—— *The City of God*, ed. Vernon J. Bourke (Garden City, NY: Doubleday Image Books, 1958).

AYER, A. J., *Language, Truth and Logic*, rev. edn. (New York: Dover Books, 1946).

BARASH, DAVID, *The Whisperings Within* (Harmondsworth: Penguin, 1979).

BEECHER, HENRY WARD, *Evolution and Religion* (New York: Fords, Howard and Hulbert, 1885).

BERGSON, HENRI, *Creative Evolution*, trans. Arthur Mitchell, with a Foreword by Irwin Edman (New York: Random House Modern Library, 1944; originally published in 1907).

BINDER, EANDO, 'The Teacher from Mars', in *My Best Science Fiction Story*, ed. Leo Margulies and Oscar J. Friend (New York: Pocket Books, 1954).

BLACK, MAX, 'The Gap between "is" and "Should"', *Philosophical Review*, 73 (1964). Reprinted in *The Is–Ought Question*, ed. W. D. Hudson (London: Macmillan, 1969).

BUTLER, JOSEPH, *The Works of Joseph Butler* (2 vols.; Oxford: Clarendon Press, 1896).

CARNEGIE, ANDREW, *The Gospel of Wealth* (first published in 1900). Reprinted in *Darwin: A Norton Critical Edition*, ed. Philip Appleman (New York: Norton, 1979).

[CHAMBERS, ROBERT] *Vestiges of the Natural History of Creation* (London: John Churchill, 1844). [But published anonymously.]

CHOMSKY, NOAM, *Aspects of the Theory of Syntax* (Cambridge, Mass.: MIT Press, 1965).

CLARK, RONALD W., *The Huxleys* (New York: McGraw-Hill, 1968).

—— *The Survival of Charles Darwin: A Biography of a Man and an Idea* (New York: Random House, 1984).

COBBE, FRANCIS POWER, *Darwinism in Morals and Other Essays* (London: Williams and Norgate, 1872).

—— *Life of Francis Power Cobbe* (2 vols.; Boston: Houghton, Mifflin, and Company, 1894).

COHEN, CARL, 'The Case for the Use of Animals in Biomedical Research', *The New England Journal of Medicine*, 315 (1986), 865–70.

CONNER, FREDERICK W., *Cosmic Optimism: A Study of the Interpretation of Evolution by American Poets from Emerson to Robinson* (Gainesville: University of Florida Press, 1949).

CROCKER, D. R., 'Anthropomorphism: Bad Practice, Honest Prejudice?', in *The Understanding of Animals*, ed. Georgina Ferry (Oxford: Basil Blackwell, 1984).

CUVIER, GEORGES, *The Animal Kingdom* (London: Henry G. Bohn, 1863; originally published in 1829–30).

DARWIN, CHARLES, *The Autobiography of Charles Darwin*, ed. Nora Barlow (New York: W. W. Norton, 1969).

—— *Charles Darwin's Notebooks, 1836–1844*, transcribed and ed. Paul H. Barrett *et al.* (Ithaca: Cornell University Press, 1987).

—— *The Collected Papers of Charles Darwin*, ed. Paul H. Barrett (2 vols.; Chicago, University of Chicago Press, 1977).

—— *The Correspondence of Charles Darwin*, ed. Frederick Burkhardt and Sydney Smith (Cambridge: Cambridge University Press, 1985 and following). This definitive edition of Darwin's letters will replace all others. At the time of this writing (May 1989) four volumes, covering Darwin's correspondence through 1850, have appeared.

—— *The Descent of Man, and Selection in Relation to Sex* (London: John Murray, 1871). Facsimile edition published by Princeton University Press, 1981. The second edition (1874) is conveniently reprinted in the Modern Library series by Random House, New York, n.d.

—— *The Expression of the Emotions in Man and Animals* (London: John Murray, 1872). Reprinted by the University of Chicago Press, 1965.

—— *The Formation of Vegetable Mould, through the Action of Worms* (London: John Murray, 1881). Facsimile edition published by the University of Chicago Press, 1985.

—— *Journal of Researches into the Natural History and Geology of Various Countries Visited by HMS 'Beagle'* (London: Henry Colburn, 1839; 2nd edn., 1845; final edn. revised by Darwin 1860). The 1860 edition is reprinted as *The Voyage of the Beagle*, ed. Leonard Engel (Garden City, NY: Anchor Books, 1962). Page references are to the latter edition.

—— *The Life and Letters of Charles Darwin*, ed. Francis Darwin (2 vols.; London: John Murray, 1888). Also published in a three-volume edition.

—— *Metaphysics, Materialism, and the Evolution of Mind: Early Writings of Charles Darwin*, transcribed and annotated by Paul H. Barrett; with a commentary by Howard E. Gruber (Chicago: University of Chicago Press, 1974).

—— *The Movements and Habits of Climbing Plants* (London: John Murray, 1865).

—— *On the Origin of Species by Natural Selection* (London: John Murray, 1859). Facsimile of the first edition reprinted by Harvard University Press, 1964. The second edition (1860) is conveniently reprinted in the Modern Library series by Random House, New York, n.d.

—— *On the Various Contrivances by which British and Foreign Orchids are Fertilised by Insects* (London: John Murray, 1862; 2nd edn., 1877). Second edition reprinted by the University of Chicago Press, 1984.

—— *The Structure and Distribution of Coral Reefs* (London: Smith, Elder and Company, 1842). Reprinted by the University of Arizona Press, Tucson, 1984.

—— *The Variation of Animals and Plants under Domestication* (2 vols.; London: John Murray, 1868; American edn., New York: Appleton, 1896).

—— and HENSLOW, JOHN STEVENS, *Darwin and Henslow: Letters 1831–1860*, ed. Nora Barlow (Berkeley: University of California Press, 1967).

—— and WALLACE ALFRED RUSSEL, *Evolution by Natural Selection* (Cambridge: Cambridge University Press, 1958). In addition to the materials that Darwin and Wallace published 'jointly' in 1858, this volume includes Darwin's first unpublished works on natural selection, the Sketch of 1842, and the Essay of 1844.

DARWIN, ERASMUS, *Zoonomia, or the Laws of Organic Life* (2 vols.; London: J. Johnson, 1794–6).

DENNETT, DANIEL C., *Brainstorms* (Montgomery, Vt.: Bradford Books, 1978).

DESCARTES, RENÉ, *The Philosophical Works of Descartes*, trans. Elizabeth S. Haldane and G. R. T. Ross (2 vols.; New York: Dover Books, 1955).

DEWEY, JOHN, *The Influence of Darwin on Philosophy and Other Essays in Contemporary Thought* (New York: Henry Holt and Company, 1910).

DUPREE, A. HUNTER, *Asa Gray* (Cambridge, Mass.: Harvard University Press, 1959).

EISELEY, LOREN, *Darwin and the Mysterious Mr X* (New York: Harcourt Brace Jovanovich, 1979).

ERLANGER, STEVEN, 'A Scholar and Suicide: Trying to Spare a Family', *New York Times*, 26 Oct. 1987.

FERRY, GEORGINA (ed.), *The Understanding of Animals* (Oxford: Basil Blackwell, 1984).

FLEW, ANTONY, *Evolutionary Ethics* (London: Macmillan, 1967).

FOSSEY, DIAN, *Gorillas in the Mist* (Boston: Houghton Mifflin, 1983).

FREUD, SIGMUND, *The Future of an Illusion*, trans. James Strachey (New York: W. W. Norton and Company, 1961).

GILLESPIE, NEAL C., *Charles Darwin and the Problem of Creation* (Chicago: University of Chicago Press, 1979).

GODLOVITCH, STANLEY, GODLOVITCH, ROSLIND, and HARRIS, JOHN (ed.), *Animals, Men, and Morals* (New York: Taplinger, 1972).

GOODALL, JANE (LAWICK), *In the Shadow of Man* (Glasgow: William Collins, 1971).

GOULD, STEPHEN JAY, 'Darwinism Defined: The Difference between Fact and Theory', *Discover* (Jan. 1987), 64–70.

—— *Ever Since Darwin* (New York: W. W. Norton and Company, 1977).

—— *The Flamingo's Smile* (New York: W. W. Norton and Company, 1985).

—— *Hen's Teeth and Horse's Toes* (New York: W. W. Norton and Company, 1983).

—— *The Panda's Thumb* (New York: W. W. Norton and Company, 1980).

GRAY, ASA, *Darwiniana: Essays and Reviews Pertaining to Darwinism*, ed. A. Hunter Dupree (Cambridge,: Mass. Harvard University Press, 1963).

—— *Letters of Asa Gray*, ed. Jane Loring Gray (2 vols.; Boston: Houghton, Mifflin, and Company, 1893).

—— *Natural Science and Religion: Two Lectures Delivered to the Theological School of Yale College* (New York: Charles Scribner's Sons, 1880).

GRUBER, HOWARD E., *Darwin on Man* (Chicago: University of Chicago Press, 1981).

HAECKEL, ERNST, *Generelle Morphologie* (2 vols.; Berlin: Reimer, 1866).

—— *Natürliche Schöpfungsgeschichte* (Berlin: Reimer, 1868).

HALL, EVERETT W., *Modern Science and Human Values* (New York: Dell Publishing Company, 1956).

HARLOW, H. F., DODDSWORTH, R. O., and HARLOW, M. K., 'Total Isolation in Monkeys', *Proceedings of the National Academy of Science*, 54 (1965), 90–2.

HARLOW, H. F., and HARLOW, M. K., *Lessons from Animal Behavior for the Clinician* (London: National Spastics Society, 1962).

HARLOW, H. F., and SOUMI, S. J., 'Induced Psychopathology in Monkeys', *Engineering and Science*, 33 (1970), 8–14.

HOBBES, THOMAS, *Leviathan*, ed. Michael Oakeshott (Oxford: Basil Blackwell, 1960; first published in 1651).

HOFSTADTER, RICHARD, *Social Darwinism in American Thought*, rev. edn. (Boston: Beacon Press, 1955).

HOOKER, J. D., *Life and Letters of Sir Joseph Dalton Hooker* (2 vols.; London: John Murray, 1918).

HULL, DAVID L., *Darwin and His Critics* (Chicago: University of Chicago Press, 1973).

HUME, DAVID, *Dialogues Concerning Natural Religion* (New York: Hafner, 1957; originally published in 1779).

—— *A Treatise of Human Nature*, ed. L. A. Selby-Bigge (Oxford: Oxford University Press, 1888).

HUTTON, JAMES, *Theory of the Earth* (Edinburgh, 1795).

HUXLEY, T. H., *Evidence as to Man's Place in Nature* (London: Williams and Norgate, 1863). Reprinted by the University of Michigan Press, Ann Arbor, 1959).

—— *Science and Christian Tradition* (New York: Appleton and Company, 1897).

—— *Science and Culture and Other Essays* (New York: Appleton and Company, 1888).

KANT, IMMANUEL, *Foundations of the Metaphysics of Morals*, Trans.

Lewis White Beck (Indianapolis: Bobbs-Merrill, 1959).

—— *Lectures on Ethics*, Trans. Louis Infield (New York: Harper and Row, 1963).

KITCHER, PHILIP, *Vaulting Ambition* (Cambridge, Mass.: MIT Press, 1985).

KNIGHT NEWS SERVICE, 'Doctor Who Told How He Helped Terminally Ill Wife is Indicted', *Birmingham News*, 10 Sept. 1987.

KUHSE, HELGA, *The Sanctity-of-Life Doctrine in Medicine* (Oxford: Oxford University Press, 1987).

LAIDLER, KEITH, 'Language in the Orang-utan', in *Action, Gesture, and Symbol: The Emergence of Language*, ed. Andrew Lock (London: Academic Press, 1978).

—— *The Talking Ape* (London: Collins, 1980).

LAMARCK, JEAN BAPTISTE, *Zoological Philosophy* (Chicago: University of Chicago Press, 1984; originally published in 1809).

LORENZ, KONRAD, *King Solomon's Ring*, trans. M.K. Wilson (New York: Crowell, 1952).

—— *On Aggression* (New York: Harcourt Brace Jovanovich, 1966).

LUCAS, ERHARD, 'Marx und Engels: Auseinandersetzung mit Darwin zur Differenz zwischen Marx und Engels', *International Review of Social History*, 9 (1964), 433–69.

LYELL, CHARLES, *The Geological Evidences of the Antiquity of Man* (London: John Murray, 1863).

—— *Life, Letters, and Journals of Sir Charles Lyell* ed. Katherine Murray Lyell (2 vols.; London: John Murray, 1881).

—— *Principles of Geology* (3 vols.; London: John Murray, 1830–3).

MACKINNON, JOHN, 'The Behavior and Ecology of Wild Orang-Utans', *Animal Behavior*, 22 (1974), 3–74.

—— *In Search of the Red Ape* (London: Collins, 1974).

MALTHUS, THOMAS, *An Essay on the Principle of Population*, 5th edn. (3 vols.; London: John Murray, 1817).

MASSERMAN, JULES H., WECHKIN, STANLEY, and TERRIS, WILLIAM, '"Altruistic" Behavior in Rhesus Monkeys', *American Journal of Psychiatry*, 121 (1964), 584–5.

MAVRODES, GEORGE, I., '"Creation Science" and Evolution' (letter), *The Chronicle of Higher Education*, 7 Jan. 1987, 43.

MAYR, ERNST, *The Growth of Biological Thought* (Cambridge, Mass.: Harvard University Press, 1982).

MOORE, G. E., *Principia Ethica* (Cambridge: Cambridge University Press, 1903).

MORRIS, DESMOND, *The Naked Ape* (New York: McGraw-Hill, 1967).

NAGEL, ERNEST, *The Structure of Science* (New York: Harcourt, Brace, and World, 1961).

NAGEL, THOMAS, 'Ethics as an Autonomous Theoretical Subject', in *Morality as a Biological Phenomenon*, ed. Gunther S. Stent (Berkeley: University of California Press, 1978).

NOZICK, ROBERT, 'About Mammals and People', *The New York Times Book Review*, 27 Nov. 1983.

NOZICK, ROBERT, *Anarchy, State and Utopia* (New York: Basic Books, 1974).

PALEY, WILLIAM, *Evidences of the Existence and Attributes of the Deity, Collected from the Appearances of Nature* (London: Faulder, 1802). Partially reprinted in *A Modern Introduction to Philosophy*, ed. Paul Edwards and Arthur Pap, 3rd edn. (New York: Free Press, 1973).

PASSMORE, JOHN, *Man's Responsibility for Nature* (New York: Charles Scribner's Sons, 1974).

PERRY, RALPH BARTON, *The Thought and Character of William James* (2 vols.; Boston: Little, Brown, and Co., 1935).

PIUS XII, POPE, *Humani Generis* (False Trends in Modern Teaching), trans. R. A. Knox (London: Catholic Truth Society, 1953).

RAWLS, JOHN, *A Theory of Justice* (Cambridge, Mass.: Harvard University Press, 1971).

REGAN, TOM, *The Case for Animal Rights* (Berkeley: University of California Press, 1983).

—— and SINGER, PETER (eds.), *Animals Rights and Human Obligations* (Englewood Cliffs: Prentice-Hall, 1976).

RICKABY, JOSEPH, *Moral Philosophy* (London, 1892).

SEDGWICK, ADAM, *The Life and Letters of the Reverend Adam Sedgwick*, ed. John Willis Clark and Thomas McKenny Hughes (2 vols.; Cambridge: Cambridge University Press, 1890).

SINGER, PETER, *Animal Liberation* (New York: New York Review Books, 1975).

SKINNER, B. F., 'Behaviorism at Fifty', in *Behaviorism and Phenomenology*, ed. T. W. Wann (Chicago: University of Chicago Press, 1964).

SOUMI, STEPHEN J., and HARLOW, HARRY F., 'Depressive Behavior in Young Monkeys Subjected to Vertical Chamber Confinement', *Journal of Comparative and Physiological Psychology*, 80 (1972), 11–13.

SPENCER, HERBERT, *The Data of Ethics* (New York: Thomas Y. Crowell & Company, 1879).

—— *Social Statics* (London: John Chapman, 1851).

TIGER, LIONEL, *Men in Groups* (New York: Random House, 1969).

TRIVERS, R., and HARE, H., 'Haplodiploidy and the Evolution of the Social Insects', *Science*, 191 (1976), 249–63.

VAN DEN BERGHE, PIERRE, *Human Family Systems* (New York: Elsevier North-Holland, 1979).

WALLACE, ALFRED RUSSEL, *Alfred Russel Wallace: Letters and Reminiscences*, ed. James Marchant (2 vols.; London: Cassell and Company, 1916).

WECHKIN, STANLEY, MASSERMAN, JULES H., and TERRIS, WILLIAM, JR., 'Shock to a Conspecific as an Aversive Stimulus', *Psychonomic Science*, 1 (1964), 47–8.

WESTFALL, RICHARD S., *The Construction of Modern Science: Mechanisms and Mechanics* (Cambridge: Cambridge University Press, 1971).

WILSON, EDWARD O., 'Human Decency is Animal', *New York Times Magazine*, 12 Oct. 1975.

—— *Sociobiology: The New Synthesis* (Cambridge, Mass.: Harvard University Press, 1975).

WITTGENSTEIN, LUDWIG, *Tractatus Logico- Philosophicus,* trans. D. F. Pears and B. F. McGuiness (London: Routledge & Kegan Paul, 1961).

WOOLDRIDGE, DEAN E., *The Machinery of the Brain* (New York: McGraw-Hill, 1963).

ZIRKLE, CONWAY, *Evolution, Marxian Biology, and the Social Scene* (Philadelphia: University of Pennsylvania Press, 1959).

INDEX